Preface

The use of bioassay procedures for assessing internal radionuclide deposition began in the 1930s. However, in the early years, the use of bioassay procedures was hampered by the instrumental sensitivity and analytical techniques available. As late as 1960, the Federal Radiation Council expressed its concern about the lack of metabolic data for humans. The development of large-volume scintillation detectors and multi-channel analyzers permitted the rapid analysis of gamma-ray emitters where only a few radionuclides were involved. More recently the development of very high resolution solid state detectors, multi-channel analyzers, computer systems, and metabolic models for the reference man have made possible the routine use of bioassay procedures for assessing internal deposition of radionuclides.

However, frequently unanswered are questions such as what bioassay is, why it should be done, on whom should it be done, ways in which it can be done, how often it should be done, and how the results might be interpreted. This report addresses these questions and is especially meant for the individual who has not been intimately involved with bioassay and desires a better understanding of the use of these procedures in radiation protection programs.

The report recognizes the limitations of bioassay procedures as a means of estimating the dose equivalent from internally deposited radionuclides. Estimates of radionuclide intakes or body (organ) burdens and their associated dose equivalents are limited by errors in the measurements made, uncertainties in the models used to interpret the results, and variability among subjects in radionuclide metabolism and dosimetry. These problems are addressed and references are provided for additional information. References are also included which cover individual radionuclide bioassay procedures.

During the period through 1986 measurements will be shown in conventional units followed by the value in SI units in parentheses (see NCRP Report No. 82).

This report was prepared by Scientific Committee 54 on Bioassay for Assessment of Control of Intake of Radionuclides. Serving on the Scientific Committee were:

Harry F. Schulte, *Chairman*
Albuquerque, New Mexico

Robert E. Alexander
U.S. Nuclear Regulatory Commission
Washington, DC

Bruce B. Boecker
Lovelace Inhalation Toxicology Research
  Institute
Albuquerque, New Mexico

Orval L. Olson
University of Missouri
Columbia, Missouri

Ralph G. Smith
University of Michigan
Ann Arbor, Michigan

Lloyd Tepper
Air Products and Chemicals, Inc.
Allentown, Pennsylvania

Consultant
William D. Moss
Health, Safety and Environment Division
Los Alamos National Laboratory
Los Alamos, New Mexico

NCRP Secretariat—**Constantine J. Maletskos**
E. Ivan White

The Council wishes to express its appreciation to the committee members and consultant for the time and effort devoted to the preparation of this report.

Warren K. Sinclair
President, NCRP

Bethesda, Maryland
December 15, 1986

NCRP REPORT No. 87

Use of Bioassay Procedures for Assessment of Internal Radionuclide Deposition

Recommendations of the
NATIONAL COUNCIL ON RADIATION
PROTECTION AND MEASUREMENTS

Issued February 28, 1987

National Council on Radiation Protection and Measurements
7910 WOODMONT AVENUE / BETHESDA, MD. 20814

LEGAL NOTICE

This report was prepared by the National Council on Radiation Protection and Measurements (NCRP). The Council strives to provide accurate, complete and useful information in its reports. However, neither the NCRP, the members of NCRP, other persons contributing to or assisting in the preparation of this report, nor any person acting on the behalf of any of these parties (a) makes any warranty or representation, express or implied, with respect to the accuracy, completeness or usefulness of the information contained in this report, or that the use of any information, method or process disclosed in this report may not infringe on privately owned rights; or (b) assumes any liability with respect to the use of, or for damages resulting from the use of, any information, method or process disclosed in this report, *under the Civil Rights Act of 1964, Section 701 et seq. as amended 42 U.S.C. Section 2000e et seq. (Title VII) or any other statutory or common law theory governing liability.*

Library of Congress Cataloging-in-Publication Data

National Council on Radiation Protection and Measurements.
  Use of bioassay procedures for assessment of internal radionuclide deposition.

  (NCRP report ; no. 87)
  "Issued December 30, 1986.
  Includes Index.
  1. Radioisotopes in the body—Measurement. 2. Biological assay. I. Title. II. Series.
  QP82.2.R3N35 1986   612'.014486   86-33136
  ISBN 0-913392-83-9

Copyright © National Council on Radiation
Protection and Measurements 1987

All rights reserved. This publication is protected by copyright. No part of this publication may be reproduced in any form or by any means, including photocopying, or utilized by any information storage and retrieval system without written permission from the copyright owner, except for brief quotation in critical articles or reviews.

Contents

Preface	iii
1. Introduction	1
1.1 Scope	1
1.2 Definition of Bioassay	1
1.3 Uses of Bioassays	2
1.3.1. Personnel Evaluation	2
1.3.2. General Exposure Control	3
1.3.3. Improved Metabolic Data	3
1.4 Medical Uses of Bioassay	4
1.4.1. General Monitoring Results	4
1.4.2. Significant Exposures	4
1.5 A Strategy of Bioassay	6
2. Necessity for Bioassay	7
2.1 Exposure Assessment	7
2.2 Radiotoxicity Assessment	9
3. Participation in a Routine Bioassay Program	11
4. Bioassay Frequency	13
4.1 Chronic Exposure Conditions	13
4.2 Single Exposure Conditions	17
4.3 Bioassays Related to Use of Respiratory Protective Equipment	18
5. Bioassay Techniques	19
5.1 Radionuclide Biokinetics	19
5.2 *In-vivo* Measurements	21
5.3 *In-vitro* Measurements	25
5.3.1 Urine Analysis	25
5.3.2 Feces Analysis	28
5.3.3 Blood Analysis	29
5.3.4 Breath Analysis	29
5.3.5 Other Biological Analyses	30
5.3.6 Collection of *In-vitro* Samples	31
6. Interpretation of Bioassay Results	32
6.1 General Considerations	32
6.2 Examples of Bioassay Interpretations	33
6.2.1 *In-vivo* Measurements of ^{134}Cs	33

Contents

      6.2.2 *In-vitro* Measurements of ^{134}Cs 37
  6.3 Other Radionuclides 44
7. Action Points and Action Based on Bioassay Results .. 46
  7.1 Preparatory Evaluation 46
  7.2 Exposure Control 46
  7.3 Diagnostic Evaluation of Bioassay Measurements 49
8. Perspective on Bioassay 50
  8.1 Errors in Measurement 50
  8.2 Uncertainties in Internal Dose Assessment 51
  8.3 Adequacy of Interpretation of Bioassay Results 54
Appendix A: Glossary 55
References 59
The NCRP 66
NCRP Publications 73
Index 81

I. Introduction

1.1 Scope

This report is concerned chiefly with the conceptual framework and methodology required for proper assessments of the occurrences, magnitudes, and retentions of internal depositions of radionuclides in people. Students, health protection personnel, occupational health personnel, and managers who may have had little or no experience in bioassay activities are likely to benefit from reading this report. Therefore, this report has been written as a basic text to acquaint the reader with the various elements of a successful bioassay program and their limitations for health protection purposes.

General information on bioassay procedures is given, as is guidance on critical elements of a bioassay program such as when bioassay is needed, who should participate, how frequently sampling should occur, how the results might be interpreted, and the limitations of bioassay data for health protection purposes. Basic principles are stressed along with general guidance for their application.

Recognizing that bioassay procedures, models, and interpretations can differ for each radionuclide, no attempt has been made to incorporate radionuclide-specific recommendations or models in this report other than by reference. It is expected that specific recommendations of this type will be published in the future by the NCRP.

1.2 Definition of Bioassay

For the purpose of this report, bioassay is the determination of the kind, quantity, location, and/or retention of radionuclides in the body by direct (*in vivo*) measurement or by *in vitro* analysis of material excreted or removed from the body. It might be considered the final quality control procedure that is used to assure adequate protection of workers against internal radiation exposure. Applications of bioassay are not limited to radionuclides. They are commonly found in occupational health programs dealing with metals (e.g., lead, mercury), and other industrial chemicals (*e.g.*, fluoride). Several uses of this type are discussed by Friberg (1985).

It is possible that biological responses (e.g., enzyme or chromosomal changes) to low levels of internal or external radiation may eventually be identified as having exposure-monitoring uses such that they might be employed in the "bioassay" of exposure. Such approaches, however, are presently available only for high-level exposures and are outside the definition of "bioassay" used in this report. Other usages of the term "bioassay" unrelated to this report refer to the use of biological systems (e.g., laboratory animals, tissue cultures) to determine the physiological, pharmacological, or toxicological properties of substances such as drugs and industrial chemicals.

Primary radiation protection standards and guides are expressed as maximum permissible annual dose equivalents, i.e., dose limits for the whole body, parts of the body, or certain organs and tissues. Because radiation doses delivered by internally deposited radionuclides cannot be measured directly, *secondary* standards in terms of organ or tissue burdens or intake levels have been derived. Only in exceptional cases such as biopsies, surgical specimens, or autopsy tissue can organ tissue burdens be measured directly. For this reason, measurements of radiation by external detection or by analysis of material excreted or removed from the body have been used in metabolic and dosimetric models to serve as guides for assessing organ or tissue dose equivalents. Thus, the normally used bioassay procedures for detecting an internal deposition do not involve direct measurements of either the intake, organ or tissue burdens, or doses. The reader should be aware of these limitations which are discussed further in Section 8.

1.3 Uses of Bioassays

1.3.1 *Personnel Evaluation*

The primary use of bioassay procedures is to determine whether an individual has been exposed to a radioactive material in a manner that resulted in internal deposition and, if so, to quantitate the magnitude of that deposition and its dosimetric consequences. Measurements performed prior to an individual's new job assignment, when indicated by a previous work history, give important baseline data and provide useful background information on exposures that might have occurred in past occupational assignments. Routine scheduled measurements, performed periodically after an individual is on the job, provide important input to evaluations of the extent to which the individual is adequately protected, is observing safe working practices, and is avoiding the accumulation of internally deposited radionuclides. Bioassay results may be obtained when an individual terminates a particular job assignment to document the estimated body or organ burden at that time. Bioassay procedures also have a definite role in the medical evaluation and management of potentially over-exposed individuals (for details, see Section 1.4).

1.3.2 *General Exposure Control*

Bioassay procedures provide a useful tool for evaluating the general exposure conditions throughout an operating facility. Careful analysis of serial results can indicate trends toward greater or lesser accumulations of radioactivity within the working population. These trends, in turn, may be correlated with new or altered procedures or with the use of new equipment. Bioassay results can provide information on possible exposures associated with unusual procedures for which experience is not available, and on exposures that have been occurring but were not suspected. These results can also indicate the extent to which engineered confinement and other protective measures and the air sampling program have been effective in the control of the exposures.

An increasingly important use of general exposure monitoring is to provide a legal record of exposure levels associated with various facilities and procedures. It should be recognized that adequate documentation of no- or low-exposure situations may be just as important for legal documentation as those dealing with higher exposure levels.

1.3.3 *Improved Metabolic Data*

Many guides and standards are based, in large part, on results obtained from laboratory animals exposed to radionuclides by different routes of administration. This requires appropriate extrapolations from laboratory animal exposures to human exposures. When available, data from human exposures have been invaluable in the development of standards and the validation of extrapolations of animal data to potential human exposure situations.

Although proper occupational practices are designed to minimize or eliminate the internal deposition of radionuclides in workers, it is recognized that accidental depositions can occur. Therefore, it is strongly recommended that analyzed results from individuals, exposed either in the past or the future, be published in the open literature to improve our understanding of the biokinetics of different radionuclides in the body or in individual organs and how well bioassay results correlate with organ and body burdens measured at autopsy. Only in this way can the models used in relating bioassay results to internal dose be corrected or validated.

1.4 Medical Uses of Bioassay

1.4.1 *General Monitoring Results*

The uses of bioassay for general monitoring of worker exposure are only of direct medical interest to the physician to the extent that the physician desires general information on the level and nature of workers' internal deposition of radionuclides and on the adequacy of control measures at the work place. In these applications, the characteristics of the radiation monitoring program will be determined largely by non-medical factors such as the radionuclides involved, the operations using radionuclides, previous operational and monitoring experience, the availability of laboratory support, competing assay obligations, and cost.

1.4.2 *Significant Exposures*

Specific medical uses of bioassay may arise when a significant exposure is suspected on the basis of monitoring data or as a consequence of an accident that created the potential for internal radionuclide deposition. Ideally, if such an exposure was suspected, the physician would use bioassay results and their interpretation, by a person knowledgeable in such interpretations, to:

- a. confirm the character, amount, and distribution of the internally deposited radionuclide;
- b. estimate the possible long-term organ retentions of the radionuclide and the associated dose-equivalent commitments;
- c. compare the results from a. and b. with existing standards for the radionuclide involved (organ burdens, annual intake); and
- d. appraise the need for, and efficacy of, work restrictions and/or therapeutic intervention.

Detailed guidance on the treatment of persons who have been accidentally contaminated with radionuclide is given in NCRP Report

No. 65 (NCRP, 1980). This report notes the difficulty or impossibility of following the above sequence exactly.

In the usual operational situation, as described by the NCRP (1980), the earliest information associated with an exposure event will consist of preliminary reports on what occurred, probable identification of the predominant radionuclides by history or initial spectrometry, and preliminary surveys of air and surface contamination. Persons involved in the accident will generally show no symptoms or signs unless there has been trauma. Measurements of surface contamination of the skin may give evidence of the potential for internal contamination. If airborne materials have been released, nasal swipes may give evidence of an inhalation exposure, although negative findings do not eliminate the possibility of internal deposition.

Under these circumstances, the physician is rarely justified in waiting for refined assay procedures and dose calculations. His decisions will be based on approximations of probable internal contamination, past experience with the operation, and the treatment and clinical appraisal of the relative risks and benefits of acting or not acting. In most situations, the medical decision will involve the use of relatively low-risk procedures, e.g., use of blocking agents, dilution or chelation, so that definitive bioassay data are not essential to the decision to proceed with treatment.

After the immediate first-aid phase of response to a potentially contaminating event, additional bioassay information may be derived as appropriate from whole-body counts for photon-emitting radionuclides and radioanalysis of urine, feces, or blood samples for the radionuclide or radionuclides involved. Quantitative interpretation of urine bioassay data using normal biokinetic models may be impossible if chelation therapy has already been started. In such a case, the magnitude of urinary excretion may give a qualitative indication of the usefulness of the chelation therapy and the possible merits of continued treatment.

Long-term continuation of periodic bioassay measurements following an accident, such as *in-vivo* measurements and assays of excreta or other biological samples, can provide a rational basis for determining the nature and extent of additional treatment. Analysis of the existing data and consultation with medical and health physics personnel familiar with the management of radiological accidents will normally precede a commitment to long-term chelation therapy or a more aggressive treatment like lung lavage. The physician responsible for management of the patient must recognize the limitations of bioassay as discussed in Section 8.

In the matter of work assignments and restrictions, adequate bioassay data are essential for physicians, health physicists, and managers responsible for occupational health. Such determinations may have important operational or administrative implications and are commonly of great importance to the worker, employer, and legal or regulatory participants in the matter.

1.5 A Strategy of Bioassay

The foregoing material describes the range of possible uses of bioassay. To establish and conduct a successful program, one should consider each of the elements described by the following steps:

a. Determine the need for a bioassay program (Section 2);
b. Determine who should be sampled (Section 3);
c. Determine the frequency of sampling (Section 4);
d. Select the bioassay technique(s) to be used for (i) routine operational health physics considerations, and (ii) examination of suspected or known high-exposure cases (Section 5);
e. Make the necessary collections and/or measurements;
f. Determine whether methods of interpretation of bioassay results are available and, if so, which ones should be used (Section 6);
g. Use results to estimate the degree of internal deposition that has occurred and compare this with the appropriate action levels (Sections 6 and 7);
h. When routine samples indicate that the estimated exposure levels have been higher than desired based on existing health protection criteria, institute procedural changes and/or better control methodology to reduce the chance for subsequent exposures (Sections 1.3 and 7); and
i. Occupational medical personnel should be contacted if an individual has a significant intake based on approximations of probable intakes or past experience (Section 1.4).

2. Necessity for Bioassay

When considering the need for a bioassay program, it is useful to consider both the factors that can influence the occurrence, magnitude and characteristics of a possible exposure (exposure assessment), and the factors that influence the deposition, retention, dosimetry, and biological consequences (radiotoxicity assessment) of such exposures. When considered together, these assessments represent a hazard assessment that should provide an important perspective for judging the value of bioassay compared with other methods of monitoring. There are many variables to consider in such an assessment, a number of which are quite complex and not reducible to a simple quantitative form. Factors such as those outlined in the next two subsections should be considered in such assessments.

2.1 Exposure Assessment

A. *Physical properties of the material.* Substances in the forms of liquids or coarsely divided solids do not readily escape from a controlled system to become available for intake. Gases, fumes, mists, and fine particles may escape more readily and remain suspended in the air for prolonged periods of time.

B. *Total quantity of a radionuclide in process at one time.* Increasing the quantity of material in process increases the probability that if a release were to occur, it would involve detectable amounts. Thus, processing only small quantities at any one time may minimize the need for a bioassay program if this is the only postulated route of exposure.

C. *Concentration of the radionuclide in the process.* Concentration is expressed as activity per unit mass of material in the process. If the concentration is high, the escape of even a very small quantity of material from a controlled system could result in a significant intake by the operator. Therefore, one needs to consider both the specific activity of the radionuclide and the quantity of matrix material in which it is present. Some materials, such as natural uranium, have relatively low activities per gram even in pure form.

D. *Degree of containment.* Work on an open benchtop presents a greater opportunity for significant internal contamination than work in a properly operating hood or glove box. Maintaining all operations in a glove box, with proper care in making transfers into and out of the box, can minimize the need for a bioassay program.

E. *Training and experience of each person involved in the process.* An essential element in training for work with radionuclides is learning proper methods of handling materials to avoid spills and the release and spread of contaminating materials. Such handling requires a degree of care that cannot be appreciated by the untrained, inexperienced operator. One might require more frequent bioassays of the new employee and then reduce the frequency as the employee responds to training and experience.

F. *Degree of confidence in radiation monitoring systems.* In a sense, a bioassay program is a second-line monitoring program. If the other elements of the monitoring program are ineffective or only partially effective because of poor design or unavoidable circumstances, a bioassay program may be a necessity. The question must be addressed as to whether a significant release to the operator's working environment, that results in an unexpected radionuclide intake, can occur without being detected. A thorough evaluation of the monitoring system is required in terms of the probability of an undetected release. Until the evaluation is carried out, a bioassay program should be used. If significant intakes are detected, the bioassay program should be continued.

G. *Prevailing level of surface contamination in the work area.* Detection of surface contamination is the simplest form of monitoring and is done most frequently. The presence of surface contamination in the work area may be an important indication of some failure of the control system either from poor design and operation or from improper work practices. Removable radioactive material on a surface can become resuspended in air and then inhaled, or it can be transferred to food or other materials and then ingested. A bioassay program is needed then to detect resultant intakes.

H. *Prevailing level of airborne contamination in the work area.* The presence of airborne contamination is also an indication of failure to control and confine the contaminant completely. Airborne contamination is more serious than surface contamination because it can readily enter the body by inhalation. Airborne contamination levels can be expressed in quantities that are readily related to hazard indications through secondary standards such as acceptable or derived air concentrations. Persistent findings of airborne concentration levels of 10 percent or more of such standards are strong indications of the need for a bioassay program in addition to the need for closer control of the sources of contamination.

I. *Fraction of the work day a person is exposed.* Standards for acceptable intakes of radioactive materials are based on the committed dose equivalent received. Hence, these standards represent time-weighted averages. Exposure for part of a day results in less deposition and, consequently, a smaller committed dose equivalent than exposure for a full day under the same ambient conditions. A work program or process that requires the operator to be potentially or actually exposed only occasionally will probably not require a bioassay program unless the potential for exposure is high, the exposure cannot be detected by other means, or the results are needed for legal purposes.

2.2 Radiotoxicity Assessment

A. *Nuclear properties of the material in process.* The biological effect of a radionuclide depends, in part, on its nuclear properties such as physical half-life, and the type, penetrability, and energy of radiation emitted. When equal activity levels are deposited internally, photon emitters are least hazardous and alpha emitters are most hazardous; beta emitters are intermediate in their relative hazard. Bioassay should always be considered when alpha or beta emitters are handled.

B. *Biokinetics of the radionuclide.* The biokinetics of a radionuclide after its deposition in the body determine which, and to what extent, organs and tissues are irradiated. The biokinetics, in turn, are influenced by a number of physical (e.g., particle size, surface area), chemical (e.g., elemental chemistry, chemical form, *in-vivo* solubility), and biological (e.g., age, metabolism, health status) factors.

The effective retention half-life in an organ or tissue, a key determinant of the committed dose equivalent, can be no longer than the shorter half-life of the two half-lives (biological and physical) from which it is derived (see Section 6.2). For instance, tritium, with a physical half-life of about 12 years, has a biological and effective half-life in body fluids of about 10 days. Cesium-137, with a physical half-life of about 30 years, has a biological and effective half-life of about 100 days. Conversely, ^{131}I in the thyroid has a biological half-life of 120 days, but the effective half-life is about 7.5 days because of its 8-day physical half-life. Other radionuclides, such as thorium and plutonium have long effective retention times because both the physical and biological half-lives are long.

An inherent assumption in this section is that an established bioassay procedure is available for the radionuclide in question. A bioassay program should also have established levels at which certain actions are required (Section 7). Unless such levels have been established or can be established for the radionuclide under consideration, the program is of little value for monitoring or protection purposes.

Consideration of the variables discussed in this section requires judgment based on experience. Unfortunately, there is no way to combine all of the variables into an equation to decide whether a bioassay program is required or useful. Guidance on when bioassay programs may be required, based on the quantities, relative radiotoxicities of the specific nuclides involved, and the nature of the processing of unsealed radioactive materials may be found in a number of references (Alexander, 1974; ANSI, 1978, 1983; Brodsky, 1969, 1980, 1983; Johnson, 1984, 1985; NCRP, 1978; Schmier and Kistner, 1972; USNRC, 1974, 1978, 1979, 1980, 1983). The key question in deciding whether or not to have a bioassay program is "Can it yield useful information for ensuring the protection of the worker?"

3. Participation in a Routine Bioassay Program

For routine monitoring of workers, one of the most important considerations regarding participation in a bioassay program is the degree to which radionuclides in the working area are confined and prevented from entering the atmosphere of the work place. If the concentration of radionuclides is normally very low, adequate monitoring for this route of exposure may be accomplished with limited participation. If higher airborne concentrations are normally encountered, extensive participation may be required. Thus, representative air sample results, averaged over a suitable period of time, provide information necessary to the decision on participation. In the case of work with long-lived, well-retained radionuclides, participation by all workers should be considered, at least once a year.

A period of one month is usually considered to be acceptable for time-weighted averaging. If the time-weighted monthly average[1] of air sample results is, for example, 10% or less of the maximum permissible concentration (MPC) or the Derived Air Concentration (DAC), and if the largest result used in the calculation of the average is less than 30% of the MPC or DAC, radionuclide confinement should be considered sufficiently effective as to require participation in the bioassay program only by a representative sample of the exposed personnel, if only the inhalation route of exposure is considered. This sample should be composed of the most highly exposed or potentially exposed workers within a given area and should include at least 10% of the workers who have regular job assignments in the area if the total number of workers is 100 or more. Rotation of participants should be considered for personnel who are subject to similar potential exposures. If the

---

[1] The Time-Weighted Average (*TWA*) concentration is the best estimate of the average (mean) concentration over a period such as a day, week, month, or quarter. If the air sampler is operated continuously over the whole period, the resulting measured concentration is the *TWA* for the place where the sampler is located. Usually, samples are collected during typical intervals when certain operations are conducted.

In this case, the concentration is determined by air sampling during each of these operations intervals. The *TWA* is then calculated by dividing the sum of products of the concentration, c_i, and its time interval, t_i, by the sum of the time intervals including the intervals of no operations. Thus, $TWA = \sum c_i t_i / \sum t_i$.

total is between 10 and 100 workers, there should be at least 10 participants. If the total is fewer than 10 workers, all should be considered for participation.

If the time-weighted monthly average of air sample results is very much greater than 10% of the MPC or DAC, and/or the largest result used in the calculation of the average is greater than 30% of the MPC or DAC, all personnel whose regular job assignments involve work in such areas should participate in the program.

Personnel whose duties involve only observation or supervision and who spend less than 25% of the work week in an area where bioassay is required, should participate on a limited basis. The degree of participation of such personnel depends on the circumstances and is a matter of judgment.

Each case must be evaluated individually to arrive at the appropriate values to use that correspond to the 10%, 30%, and 25%, respectively, that are used in the example.

4. Bioassay Frequency

Before an individual begins work in an area in which bioassays are required, the worker's previous exposure to radioactive material(s), and the accumulated deposition and retention, should be determined by a careful examination of the worker's exposure history and/or by baseline bioassay measurements. The frequency with which bioassay measurements are subsequently made is the most critical consideration from the viewpoint of program effectiveness and cost. It is essential to perform the measurements at a frequency sufficient to accomplish the objectives of the program, but, in most cases, it is also essential to avoid the expense and inconvenience of unnecessary bioassays.

In this discussion of bioassay frequency, it is convenient to distinguish two basic types of conditions for which the frequency considerations are somewhat different, namely, chronic exposure conditions and single exposure conditions. The discussion also includes bioassays following the use of respiratory protective equipment. On the termination of work in areas in which bioassay is required, those bioassay measurements necessary to determine the accumulation of radioactive material in the body should be made for each worker.

4.1 Chronic Exposure Conditions

Where work with low toxicity radioactive materials is performed, e.g., with uranium, the process equipment and ventilation control systems are usually not designed to provide total confinement of the materials. The design objective is to control the concentration of radioactive material(s) in air at acceptably low levels. Under these conditions, worker exposure may be chronic, but the buildup of radioactive material in the body by inhalation is limited by a combination of design and process features that keep the concentrations low and by the relatively rapid radioactive decay and/or biological elimination that removes the materials from the body. In many such working environments, a virtual equilibrium develops in which the rate of elimination from the body is approximately equal to the rate of intake into the body. However, it is important to recognize that this balance can be disturbed by unplanned exposure events such as spills, fires, or exposure by ingestion, skin absorption or wound contamination.

Where worker exposure is chronic, two types of measurements are normally performed to evaluate internal deposition of radionuclides: air samples and bioassays. A properly performed air sampling program can indicate whether chronic inhalation exposure levels are sufficiently low. Air sampling can also be used to evaluate the magnitude of unplanned inhalation exposure events. Experience has shown, however, that air sampling alone is not always an adequate indicator of either. If sufficiently large quantities of radioactive material are present in the work environment, unexpected depositions in the body can and do occur. Bioassay measurements are necessary, therefore, to detect such events and to assure that the quantity of radioactive material in the body does not exceed the recommended limits.

The frequency of such measurements is dictated by two distinct objectives. The first objective is to monitor the accumulation of radioactive material in the body. The second is to assure that significant depositions are detected, so that appropriate corrections can be instituted in the working conditions. Normally, the frequency selected to meet the second objective will suffice for meeting the first.

Monitoring radionuclide accumulation. If the level of radioactive materials in air in the working area is generally low (e.g., the time-weighted average does not exceed 10% of the permitted concentration) and if transient elevations in this level are infrequent and not great (e. g., not more than three times the chronic level), annual bioassays may be adequate. For a radionuclide such as inorganic tritium, however, which has a short effective half life, and therefore, does not accumulate in the body, the objective of monitoring an accumulation does not apply. Bioassay for such substances, if it is considered necessary for the particular circumstances, is usually done more frequently than annually to confirm the air sampling results.

If the average concentration of radioactive material in air normally exceeds 10% of the permitted concentration and/or transient elevations tend to be more than three times the chronic level, it is unlikely that annual bioassays will adequately monitor the accumulation of radioactive material in the body. Under such conditions, bioassay measurements should be conducted often enough to assure that significant depositions do not go undetected (the second objective).

These general comments must be adapted to specific working conditions. More consideration may be necessary, for example, in those situations where many radionuclides are in use at the same time such as in radiochemistry and nuclear medicine laboratories. Under those conditions, the operations involved, the need for bioassay, and its sampling frequency have to be considered for all the nuclides combined, and the resulting actions may be more conservative than those involving a single nuclide.

Assuring detection of significant internal deposition of radionuclides. If the potential exists for larger depositions in the body (i.e., average air concentrations greater than 10% of those permitted and/or transient levels greater than three times the average) that may not always be predicted by air sampling techniques, it is necessary to perform bioassay measurements more frequently to assure prompt corrective action. Such depositions will go undetected if the rate of elimination of the radioactive material reduces the body burden to a very low or undetectable level before the next bioassay is performed.

The second objective may be achieved by establishing a bioassay frequency which takes into consideration all of the following:

a. A basic protection criterion, e.g., the annual dose equivalent limit for the organ of reference. (Fractions of a basic protection criterion can be used for establishing action points as described in Section 7.);
b. A biokinetic model for the radionuclide under consideration, particularly the retention function that is applicable or is to be assumed; and
c. The lower limit of detection of the measurement technique to be used.

Based on these considerations, a frequency should be selected that will assure that any deposition exceeding the protection criterion will be detected successfully with the selected measurement technique before becoming undetectable due to the elimination of the radionuclide from the body. The more sensitive the measurement technique, the longer the allowable time between measurements (Buisman, 1985).

This concept is illustrated for uranium in a relatively soluble form in Fig. 4.1 based, solely as an example, on the material presented below. The derivation of this figure (Alexander, 1974) is based on a criterion for protection of the kidney against chemical toxicity and the following retention function for Class D uranium:

$$q(t) = f\, I_u t^{-0.5}\ (t > 1), \qquad (4.1)$$

where
$q(t)$ = the systemic burden of uranium at t days after exposure
and
I_u = initial amount of uranium reaching the blood
f = fraction of I_u that becomes systemically deposited.

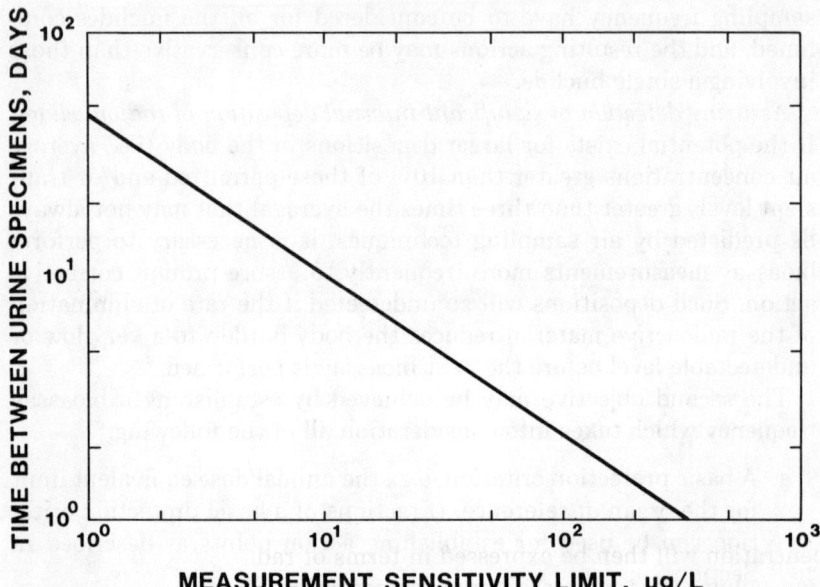

Fig. 4.1 Maximum time between urine specimen collections to detect a single uranium intake following exposure to airborne class D uranium as a function of the urinary measurement sensitivity limit (after Alexander, 1974).

An estimate of the concentration of uranium in the urine, x, is obtained by differentiating Eq. (4.1) and dividing by an estimated urinary output rate of 1.4 L/d:

$$x = \frac{0.5}{1.4} \, f \, I_u t^{-1.5} \; (t > 1). \qquad (4.2)$$

Using a value of f = 0.33 and I_u = 2700 µg (based on a nephrotoxic limit of 3 µg/g) and solving for t:

$$t = 48 \, x^{-0.67} \; (t > 1). \qquad (4.3)$$

This relationship can be used to determine the maximum time between urine specimen collections and is plotted in Fig. 4.1. Although the equation is applicable to single exposure conditions, the coefficient is based on a nephrotoxic limit for continuous exposure to uranium, namely 3 µg U/g kidney.[2]

The value of t can be interpreted as the maximum time that can occur between urine collections without losing the ability to detect a

---

[2] There are indications that a nephrotoxic limit of 3 µg U/g kidney may be too high (Morrow *et al.*, 1982). This value is used here only for illustrative purposes.

single intake that will produce a transient kidney concentration of 3 µg/g. For example, if the measurement sensitivity in Fig. 4.1 is 6 µg/L, specimens must be collected every two weeks. However, if the sensitivity is 2 µg/L, a value attainable with some extra care, then specimens can be obtained monthly. Note that in the case of uranium and other naturally occurring radionuclides, as the sensitivity of the measurement technique increases and the detectable amount in the urine decreases, attention must be given to laboratory contamination and to natural background levels in human urine. A practical limit of one year between specimen collections has already been recommended in Section 3 for all radionuclides with long effective half-lives.

Both the form of Eq. (4.1) and the constants vary among people for different radionuclides and their associated different physical, chemical and metabolic properties. If radiotoxicity is involved, as is usually the case, rather than chemical toxicity as used in the illustration, then the protection limit becomes the dose limit expressed, for example, as the committed dose equivalent to a particular organ. The urine concentration will then be expressed in terms of radioactive material per unit volume. For some radionuclides, retention can be expressed as a power function, as in this example, or as one or more exponential terms. Whatever the retention, an expression analogous to that of Eq. (4.3) can be obtained for selecting the interval between specimen collections.

In summary, for true chronic exposure conditions where there is equilibrium between the rate of intake and the rate of elimination, any single bioassay result can be used to calculate the body burden. Where the air concentrations are low and relatively constant, an annual bioassay is adequate. Where air concentrations are relatively high and/or variable, the sampling frequency should be increased such that a significant intake is not overlooked, because the bioassay level has diminished to an undetectable level before the bioassay is performed. This is particularly important where the bioassay is done on urine samples, since a low excretion rate may still be consistent with a significant accumulation of unknown magnitude if the time between a single large intake and the collection of the bioassay sample is long. If the bioassay can be done by *in-vivo* counting, the count obtained is a direct measure of the quantity present.

4.2 Single Exposure Conditions

Where work is conducted with a very toxic radionuclide such as plutonium, total containment is normally attempted. Under these working conditions, there is little chronic exposure, and significant depositions in the body result only from accidental releases to the working environment. Radionuclides, such as plutonium, may also become airborne during maintenance operations, but planning for such operations normally specifies the use of respiratory protective equipment which, if properly used, essentially precludes significant intakes.

Accidental releases rarely go undetected. As noted earlier, monitoring equipment is usually employed for the purpose of rapid detection of any breach in the containment systems. Such breaches are normally repaired promptly. Under working conditions of this nature, there is less incentive to use bioassay measurements as a method for ensuring that exposures will not go undetected. The principal objective of routine bioassay measurements, then, should be to monitor the possible accumulation of radioactive material in the body. Where exposures are potentially high, the frequency must be greater but can only be chosen arbitrarily. After each significant accidental release, however, bioassay measurements should be made on all potentially affected personnel.

4.3 Bioassays Related to the Use of Respiratory Protective Equipment

Situations often arise in which workers must use respiratory protection devices. For example, a maintenance operation may entail potential exposure to airborne radioactive material and respirators may be worn as a precaution. In other situations, the operation may involve disassembly, cutting, grinding, or other procedures that will definitely generate radioactive particles and, in such cases, respirators are primary protection devices. A third type of application involves an emergency situation, such as a fire, where it is prudent to assume that radioactive materials are airborne and that respiratory protection is needed. It is clear from these examples that personnel wearing respiratory protection may be working in areas having air that can range from being relatively clean to highly contaminated.

Respiratory protection devices do not always offer complete protection and bioassay, rather than air-concentration measurements, must be used to evaluate possible exposures. The appearance of the radionuclide in the bioassay sample of an individual who wore a respirator when exposed to high concentrations of airborne radionuclides may be an indication of the ineffectiveness of the respirator. A bioassay measurement should also be conducted if, for any reason, the magnitude of the exposure (assuming no respiratory protection device) is unknown.

5. Bioassay Techniques

5.1 Radionuclide Biokinetics

Radioactive material can gain access to the interior of the body by ingestion, by inhalation, through wounds or by absorption through the skin as indicated by the large arrows in Fig. 5.1. The subsequent internal disposition of the radioactive material, i.e., its metabolism, is determined by the physical and chemical form of the radioactive material and its associated *in-vivo* solubility. The pathways by which radionuclides are metabolized after reaching the blood are the same as those followed by stable forms of the same chemical elements.

When a radioactive material is ingested, it passes through the gastrointestinal tract mixed with the contents of the tract. During this transit, some absorption of the radioactive material may occur, and the remainder is excreted in the feces. The fraction absorbed can be very high, as in the case of soluble radionuclides such as tritium, iodine or cesium, or very low, as in the case of relatively insoluble forms of radionuclides such as plutonium or uranium (ICRP, 1979). The absorbed fraction is circulated via the bloodstream and depending on the element, it will be deposited in specific organs where metabolic factors control its deposition and its further excretion via the urine and feces.

Various fractions of inhaled material, depending on particle size, are deposited in different regions of the respiratory tract, and the fraction not deposited is exhaled. The regional deposition of the inhaled material is influenced by the characteristics of the inhaled material (chiefly size) and by the anatomic and physiologic status of the subject. Once deposited, the elimination and systemic absorption of the material are governed by the sites of deposition and by the physical and chemical properties of the inhaled particles. For instance, particles deposited initially on the ciliated epithelium of the respiratory tract are cleared rapidly by mucociliary activity followed by swallowing and passage through the gastrointestinal tract. If the inhaled material is soluble in body fluids, significant absorption to the bloodstream can occur through the tissue of the upper respiratory tract or gastrointestinal tract during this clearance process.

Retention of material deposited initially beyond the level of the ciliated epithelium (i.e., below the terminal bronchioles) is strongly influenced by the solubility of the material in the body fluids. Soluble materials (e.g., ^{137}CsCl or ^{90}SrCl$_2$) are quickly absorbed into the bloodstream (Boecker, 1969; McClellan et al., 1972). Relatively insoluble forms are retained in the pulmonary region for longer times and are subject to removal by both mechanical and dissolution-absorption processes. Movement of slowly or poorly transferable[3] or non-transferable material (e.g., UO$_2$, PuO$_2$) in particulate form to the pulmonary lymph nodes represents another important mechanical clearance process (Thomas, 1968). Factors that can influence the *in-vivo* solubility of an internally deposited radionuclide include its elemental chemistry, chemical form, surface area, and specific activity.

Solubility in body fluids is also an important factor in the direct absorption of material through the skin. In the case of a puncture wound, solubility influences the rate at which the deposited material will leave the wound site via the lymph and enter the circulation for distribution to various organs.

Bioassay procedures are used to estimate the body burden of a radioactive material and the distribution among different organs that may follow internal deposition via any route of exposure. Because the direct analysis of tissue samples is seldom possible in living persons, other methods must be used. The method of choice depends on the ease of detection of the emissions from the internally deposited material, the distribution of the material within the body, and the possible changes in distribution with time after exposure.

In-vivo measurement, in which emissions from the deposited radioactive material are detected externally to the body, is the most direct method of quantifying internally deposited radioactive materials. However, not all radionuclides emit radiations that are sufficiently penetrating to be detected outside the body. For bioassay of such radionuclides indirect methods must be used (Gautier, 1983). These methods involve the collection and analysis of *in-vitro* samples that are excreted, secreted, or removed from the body (urine, feces, blood, breath, hair, perspiration), and involve the inference of total body and/or organ burdens by the use of a metabolic model. These common samples are shown as circles in Fig. 5.1. On special occasions, biopsy samples of particular tissues may be necessary, such as bone or tissue from a wound.

---

[3] The word "transfer" and its derivatives are used as general, non-specific, terms to describe overall movement of dissolved radioactive material from one place in the body to another without implying any particular mechanism. In common usage, when the rate of movement is fast, reference is made to transferable or rapidly transferable material. When the rate is slow, reference is made to slowly or poorly transferable or non-transferable material.

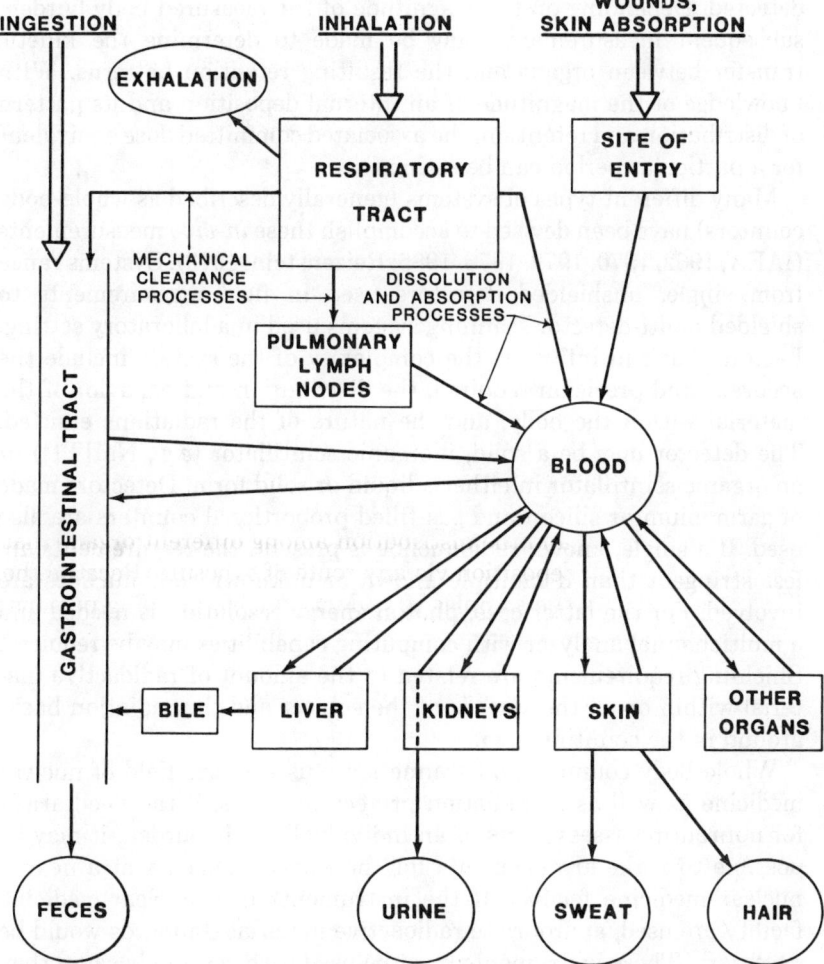

Fig. 5.1 Schematic representation of routes of entry, metabolic pathways and possible bioassay samples for internally deposited radionuclides.

5.2 *In-vivo* Measurements

In-vivo measurements are those in which the emission of photons from internally deposited radionuclides is detected external to the body. *In-vivo* measurement is the only direct estimate of body burden in a living subject. *In-vivo* measurements may be made for several different reasons, the most common being routine surveillance of workers to detect possible unknown exposures and to measure the quantities in the body and, possibly, organs when such an exposure is detected. Depending on the magnitude of the measured body burden, subsequent measurements may be made to determine the kinetic transfer between organs and the resulting retention patterns. With knowledge of the magnitude of an internal deposition and its pattern of distribution and retention, the associated committed dose equivalent for a particular period can be estimated.

Many different types of systems (generally described as whole-body counters) have been devised to accomplish these *in-vivo* measurements (IAEA, 1962, 1970, 1972, 1976, 1985; Reizenstein, 1973). Systems range from single, unshielded detectors used in field measurements to shielded multi-detector scanning systems used in a laboratory setting. Factors that can influence the complexity of the system include the accuracy and precision required, the distribution and behavior of the material within the body, and the nature of the radiations emitted. The detector may be a solid, inorganic scintillator (*e.g.*, NaI[Tl]) or an organic scintillator in either a liquid or solid form. Detectors made of germanium or silicon and gas-filled proportional counters are also used. If a single, known radionuclide is present, the requirements are less stringent than if multiple known or unknown radionuclides are involved. For the latter case, photon energy resolution is needed and a multichannel analyzer with computing capabilities may be required. Shielding requirements are related to the amount of radioactive material within or on the surface of the subject and the radiation background in the counting area.

Whole-body counters and scanners are used in the field of nuclear medicine as well as in radiation protection. Thus, if the need arises for nonroutine assessments of an individual's body burden, it may be possible to make arrangements for these measurements at a nearby nuclear medicine facility. If the instruments in a nuclear-medicine facility are used, appropriate radioactive material standards would be necessary. These instruments must be used with caution because they are generally designed with collimating and operating characteristics to maximize detection of high activity levels of low-energy gamma emitters. Information on whole-body counters throughout the world was published by the International Atomic Energy Agency (IAEA, 1970), with additional information in later publications (IAEA, 1972, 1985).

Field measurements for radionuclides such as ^{137}Cs or ^{131}I with relatively energetic photons can be made using a single, unshielded NaI detector (Palmer *et al.*, 1976) and by survey instruments. Such a device might be used to evaluate an accident to determine if there had been internal deposition of such radionuclides. Because of the simplicity of the system, it can be assembled from components available in many nuclear facilities. It can also be operated by individuals who are not highly trained. Because accuracy is not a primary feature of this system, and the potential for surface contamination of the individual is considerable, individuals who appear to have significant levels of deposited radioactivity should be counted again with more suitable equipment.

Routine *in-vivo* counting for a radiation-protection program is usually conducted within a room heavily shielded with lead or steel. A single large NaI crystal positioned at a short distance from a subject sitting in a chair or lying on a reclining chair or on a curved surface will produce reliable, easily calibrated measurements of the total body radionuclide content. Multiple, fixed-position NaI detectors provide more sensitivity and geometry independence than a single crystal while still maintaining the resolution inherent in NaI detectors. Obviously, the initial cost and the maintenance needed for the electronics increases considerably with the number of detectors used.

A useful, and sometimes less expensive, alternative to a large array of crystals involves moving one or several crystals along the length of the subject (or moving the subject with respect to a fixed detector) and determining the observed activity as a function of position of the detector. The summation of the activity recorded is a measure of total body burden, while the shape of the activity profile is a rough indication of the distribution of the radionuclide within the subject. One advantage of moving the subject instead of the detectors is that a small shadow shield can be built around the detectors, thus minimizing the size, cost, and weight of the entire system (Palmer and Roesch, 1965; Boddy *et al.*, 1975). Among other features, such a system can be built in a truck and transported to facilities requiring these services for routine checks or to the scene of an accident (Brady and Swanberg, 1965). A simplified shielded chair and crystal arrangement can also be used (e.g., Masse and Bolton, 1970). The latter arrangement provides shorter counting times and ease of entrance and exit. Also the chair is half the size of the scanning bed and can be easily adjusted for thorax counting.

Measurement of the body burden can also be made using liquid or solid organic scintillators. In these cases, the detector is usually fabricated in the shape of a hollow cylinder and the subject is placed on the midline inside the chamber. Such a configuration provides high counting efficiencies because of the approximate 4π steradian geometry. Although such devices are very sensitive for whole-body counting, they do not provide information on the sites of deposition within the body nor do they have good energy resolution.

Other strategies are sometimes employed to quantify deposition and retention in specific body organs. For instance, after inhalation of a radionuclide in a form that is relatively insoluble in body fluids, most of the radionuclide remaining in the body will be in the pulmonary region following the initial rapid clearance (1 to 2 days) of the ciliated airways (nasopharyngeal and tracheobronchial regions). If the radionuclide has sufficiently energetic emissions, it can be detected by conventional NaI (Tl) detectors. However, for a radionuclide such as ^{239}Pu, which mainly emits low energy x rays in addition to alpha particles, a more sensitive photon detector is required relative to background. Three current approaches to this problem are gas-proportional, phoswich and hyperpure germanium detectors. A hyperpure germanium detector provides better energy resolution than the other systems. Multi-detector arrays of hyperpure germanium detectors have been successfully used to quantitate lung burdens of actinides such as ^{238}Pu, ^{239}Pu, ^{241}Am, ^{235}U and ^{238}U (IAEA, 1985).

Thyroidal uptake of ^{131}I or ^{125}I is readily monitored by an external NaI(Tl) probe. Deposition of some radionuclides in the skeleton can be assessed by measuring the activity in a limb or in the skull. For weak, low-energy photon emitters such as ^{210}Pb-radiation, skull counting is useful because the overlying tissue layer is thin, the large radius of curvature permits close approach by the detector, the large surface area presented permits use of large diameter detectors, and a large fraction of the skeletal mass is associated with the skull (Wrenn et al., 1972; Olson, 1972).

If a radionuclide enters the body through a puncture wound, its early disposition depends on the form of the material. If it enters in a relatively soluble form, it leaves the wound site rapidly and is distributed throughout the body in the pattern dictated by its chemistry. On the other hand, if the injected material is relatively insoluble in body fluids, it would tend to remain at or near the wound site. If permitted to remain at a wound site, a large fraction of a sparingly soluble radionuclide will eventually be absorbed. Special wound detectors have been devised to monitor for the presence of transuranic radionuclides with low-energy photons (Epstein and Johanson, 1966; Fromhein et al., 1976).

After the inhalation and deposition of relatively insoluble materials such as UO_2 or PuO_2 in the pulmonary region, one clearance pathway leads to the tracheobronchial lymph nodes, where the material may be retained for a long period. As a means of observing this localized concentration in individuals who have had long-retained chest burdens of insoluble ^{239}PuO$_2$, an experimental, small scintillation probe was designed that could be placed in the esophagus to obtain measurements from the nearby tracheobronchial lymph nodes (Swinth et al., 1974).

Although it was a useful device for the detection of very localized radioactivity in the body, calibration of the device to allow meaningful interpretation of the results was extremely difficult because of the biological and geometrical factors involved.

5.3 *In-vitro* Measurements

5.3.1 *Urine Analysis*

Analysis for radioactive material or chemical analysis for materials such as uranium excreted in the urine may provide a useful assessment of the existing systemic burden. Urine analysis can be particularly useful when equipment and facilities for external detection of radionuclides are not available on a routine basis. Also, because some radionuclides, such as ^{3}H or ^{14}C, emit very weak beta particles without accompanying gamma photons, external detection is not possible.

When a radionuclide reaches the circulation in ionic or complexed form, fractions are deposited in different organs and filtered by the kidneys depending on the chemistry of the radionuclide. This fractionation is subject to a dynamic equilibrium based on the chemistry of the radionuclide and the cellular dynamics and biochemistry of the target organs. The radionuclide subsequently released from these organs by cellular turnover or other kinetic processes, re-enters the bloodstream where it again undergoes fractionation (including possible translocation to another body organ) with continuing excretion in the urine and feces. Thus, one can derive an indirect assessment of the amount remaining in the body from measurements of the amounts of a radionuclide excreted from the body (Jackson and Dolphin, 1966; ICRP, 1968, 1971).

Periodic analysis of urine can serve several purposes. When performed routinely, it can indicate an internal exposure that may not have been noted through an air sampling program. By providing repetitive estimates of the body burdens being accumulated in the working population, these analyses can be used to assess the performance of the radiation protection control practices. Also, in the case of an individual with a known body burden of a radionuclide, sequential analyses of urinary excretion may provide the basis for estimating the fraction of the body burden that was deposited in a transferable form.

Although various kinetic models relate urinary excretion to the amount of a radionuclide in the body, the actual data on which most of these models are based exhibit considerable variability among samples from a single individual or from different individuals (Pochin, 1968; Snyder et al., 1972). Thus, a single spot sample taken from a person usually has a high degree of physiologically related uncertainty and does not provide a reliable estimate of body burden, even with the care in collection and the applied corrections discussed in this subsection (also see Section 8).

In addition to sample variability, other problems are encountered in the proper evaluation of urinary excretion data. Since the method is an indirect assessment of body burden, one must adopt a model that describes the expected behavior of the material in the body and the temporal excretion relationships. One limitation of urine analysis is the degree to which the behavior of the material in the subject being tested matches that incorporated in the model being used. Discrepancies of this sort can be caused by differences in the subjects, the forms of the material deposited, and the mode of intake (see Section 8).

Because of the numerous assumptions that must be made for an indirect assessment, urinalysis is more successful in dealing with a radionuclide, such as ^{3}H or ^{137}Cs, that is distributed nearly uniformly throughout the body and for which the long-term retention can be adequately described, for practical purposes, by a single exponential function. In such cases, the urinary excretion should represent a constant fraction of the total existing body burden regardless of the elapsed time after exposure. For other radioelements, the urinary excretion patterns are best represented by multiple exponential functions, power functions or their combinations, and it is necessary to know the time elapsed between exposure and sample collection to relate urinary excretion to body burden. The time after a single exposure may be known. When the subject has accumulated multiple depositions, however, the effective time after exposure is difficult to determine in relation to the model used. Interpretation of the data is also complicated when the individual is, or has been, undergoing chelation therapy for removal of the deposited material (Hall et al., 1978) because the levels of radionuclide excretion in the urine will be elevated over those observed without chelation.

Practical considerations in the collection of urine samples should be noted. In many dosimetric evaluations, the quantity of radioactive material excreted per 24 hours is required and is obtained by collecting a 24-hour urine sample. For a routine bioassay program, collection of 24-hour samples may be difficult because collection is required both at home and during working hours when the possibility of sample contamination is high. Samples collected just before retiring at night

5.3 IN-VITRO MEASUREMENTS / 27

and all samples until and including the first voiding after rising in the morning on two successive days will approximate the volume of a 24-hour sample, and such samples can be suitable for many situations. Such samples, sometimes called incremental samples, represent the integral excretion over the collection period, and account must be taken of this if the excretion rate is changing rapidly during this period.

For some materials, such as natural uranium, 24-hour samples are not essential for control purposes. Analysis of a single voiding will give adequate evidence of exposure or lack thereof. In this case, consideration must be given to the time of sample collection. A sample collected at the end of the work shift on Friday is the most sensitive indicator of exposure. On the other hand, analysis of a sample taken on Monday morning before starting work is a better indicator of retained or accumulated material. A sample taken after a period of no exposure or after a vacation will indicate the presence of a burden that is cleared slowly, because the rapidly cleared, recently deposited burden will have been excreted during the no-exposure period. The purpose of the bioassay program will dictate the choice among these schedules.

Except when true 24-hour samples are collected, some correction of the concentration measurement may be required to account for abnormal conditions of high or low fluid intake or excessive loss of water by perspiration. This correction is frequently made by relating the specific gravity (sp. gr.) of the sample to the average sp. gr. of urine which is 1.024 g/mL. The correction to be applied is determined from the relation (NIOSH, 1974):

corrected concentration

$$= \text{(measured concentration)} \frac{1.024-1}{\text{(measured sp. gr. -1)}}.$$

An alternative correction may be made based on the fact that creatinine is excreted at an average rate of 1.7 g/d for men and 1.0 g/d for women (Jackson, 1966). The ratio of the expected creatinine content to the measured creatinine content of the sample provides a correction to convert the amount of radioactive material in the sample to the equivalent of a true 24-hour collection. Sample volume for a 24-hour collection or the volume collected over a known time interval may be indicative of a need for correction.

If urine samples are to be taken to evaluate exposure or dose following an accidental exposure, a urine sample should be voided immediately following the exposure and cleanup, if possible. This action avoids dilution of the subsequent sample by urine already present in the bladder at the time of the incident. This immediately voided sample may also be analyzed to provide baseline information.

5.3.2 *Feces Analysis*

Feces analysis is another means of obtaining an indirect assessment of the body burden of an internally deposited radionuclide. In addition, fecal to urinary ratios can be useful, especially if they are not constant with time after exposure. In the latter case, the value of the feces to urine ratio may give a rough idea of when the exposure occurred. Many of the principles discussed above for urine analysis apply also for feces analysis. As indicated in Fig. 5.1, radionuclides can be transferred to the feces via several different pathways, and it is important to understand these pathways in order to interpret the data.

If a radionuclide is ingested, the feces will contain the fraction of material that was not absorbed during passage through the gastrointestinal tract. If a radionuclide gains access to the systemic circulation, a fraction of the circulated material may appear in the feces via endogenous excretion into the gastrointestinal tract. Finally, if the radionuclide is deposited in the body by inhalation, a portion of the inhaled radionuclide will enter the gastrointestinal tract by mechanical clearance via mucociliary action and subsequent swallowing. The amount of material cleared from the lung by mechanical processes and appearing in the feces will be reduced by any subsequent absorption that occurs during passage through the gastrointestinal tract.

Feces analysis can be useful in detecting and quantifying an inhalation exposure to a relatively insoluble form of radionuclide, because clearance via the feces is the predominant excretion mode in such a situation (Eakins and Morgan, 1964). In the early post-exposure period (1 to 2d) the amount of a radionuclide excreted in the feces will be substantially higher than at later times due to the rapid ciliary clearance of the nasopharyngeal and tracheobronchial regions. Subsequent to this clearance, the pulmonary region controls the rate of release of a relatively insoluble radionuclide.

If feces analysis is used, several problems must be addressed. As noted above, the sources of radioactive material in the feces are more difficult to ascertain and quantify than are those for urine. Also, the collection and analysis of feces samples may be more difficult than other bioassay procedures. Appropriate fecal sampling kits (such as portable camping toilets) and methods should result in satisfactory cooperation by the user. In the case of gamma-emitting radionuclides, the objectionable aspects of processing such samples can be avoided by counting the whole sample in its closed collection container. A low-energy photon detector for the radioassay of Pu or Am in biological samples is described by Guilmette (1986).

Interpretation of the results may be more difficult than for urine samples because the daily rate of fecal mass excreted is considerably more variable than the daily rate of urinary volume (Pochin, 1968; ICRP, 1975). While the specific gravity or creatinine content of the urine can be used to adjust or normalize the results of a urinalysis, no similar index has been identified for feces. In spite of the above problems, there are circumstances when the analysis of fecal samples is the most useful bioassay method. Usually, these circumstances involve attempts to estimate initial burdens in the respiratory tract and long-term retained lung burdens by assessment of the clearance due to mechanical processes.

5.3.3 Blood Analysis

Another possible means of estimating the burden in an individual is to measure the radionuclide content of a blood sample. The information obtained would generally be comparable to that obtained from urine analysis. Blood samples show less fluctuation in radionuclide content than do urine samples, but the sensitivity of the analysis is limited, due to the amount of blood that can be withdrawn from an individual, particularly in the event of an accident with severe injury. Unfortunately, many elements are rapidly taken up by organs or bound to tissues and are only slowly released back to blood. Thus, blood bioassay, even a few hours after exposure could result in misleadingly low interpretations of blood activity concentrations. Because of these factors, as well as the ease of sampling urine, blood sampling is rarely used to monitor the possible uptake of radionuclides in individuals working with radioactive materials under normal conditions. However, in a serious exposure the level of radioactive material in the blood could be high enough to allow accurate analysis of a small sample of blood.

5.3.4 Breath Analysis

Breath samples have been analyzed to estimate the internal deposition of radium and thorium nuclides. Breath samples can also be applied to some other radioactive materials that produce a gas or vapor in the body. Radon-222, the decay product of ^{226}Ra, can be collected with air handling systems that include flasks, sampling bag, spirometers, or charcoal traps (the latter two permitting measurement of exhalation rate as well as concentration) and measured with ionization chambers or scintillation devices. Radon-220, eventual decay product of ^{228}Th or ^{232}Th, can be sampled by electrostatically collecting its ionized daughters and measuring the counting rates of these decay products with solid state detectors. Descriptions of these techniques can be found in Evans, 1935, 1937; Aub et al., 1952; Curtiss and Davis, 1943; Lucas, 1957; Vennart et al., 1964; Hickey and Campbell, 1968; Lucas et al., 1971; Harley, 1972; Keane and Brewster, 1983, and Toohey et al., 1983.

The breath of persons exposed to ^{3}H-labeled compounds contains ^{3}H$_2$O vapor. The breath of persons exposed to ^{14}C-labeled compounds contains ^{14}CO$_2$ or other volatile compounds of ^{14}C and usually can be used to assess the body burden of ^{14}C (Dancer and Finnigan, 1975; Whillans and Johnson, 1985). Except for ^{220}Rn and ^{222}Rn, however, breath is seldom analyzed, because other sample types are more readily available. While extensive measurements of ^{14}CO$_2$ and of ^{133}Xe in breath have been performed in research and diagnostic studies, these measurements are not bioassay applications. They can be used, however, for bioassay purposes if necessary and applicable.

5.3.5 Other Biological Analyses

After reaching the bloodstream, some radionuclides will eventually be incorporated in tissues that can be sampled readily. Facial and scalp hair, for example, have been shown to be a useful indicator of exposure to radon decay products in uranium miners (Savignac and Schiager, 1974). Fingernails are also integrators of exposure to some materials. Cleaning may be necessary for some of the foregoing materials, but may also remove the material wanted. Saliva may be useful on some occasions. Some of the methods previously described, when applied to the foregoing materials would have greater sensitivity and reliability. For instance, while perspiration can be analyzed to assess ^{3}H in body fluids, urine sampling would usually be a better choice. While it is easy to collect all these samples, the results obtained from them are generally qualitative indicators and cannot be interpreted quantitatively and usually do not relate to body burden.

Although biopsy or surgical specimens are not routinely used for assessing uptake of radionuclides, their analysis should not be overlooked if it is necessary to remove portions of tissue for other reasons.

Human data of this type can provide a useful check on indirect assessments made previously on the same subject. Analyses of selected tissue samples from deceased workers also provide an important evaluation of bioassay measurements made while these workers were actively involved in work with the potential for intake of radionuclides (Schulte, 1975; Heid, 1983).

5.3.6 *Collection of In-vitro Samples*

Care must be exercised in the collection of bioassay samples to prevent their contamination. Clean containers must be used for collection and storage; single-use containers can meet this requirement most readily. All biological samples are also subject to deterioration by bacteriological action that may interfere with subsequent analysis. Prompt analysis following collection is the preferred method of avoiding these complications. When samples must be kept longer than a day, they should be refrigerated, acidified to minimize precipitation, or have a preservative added to prevent bacterial growth. For some analytical techniques, it may be appropriate to add a carrier to minimize losses to the container walls or to obtain high recovery of the radioactive material. Appropriate handling procedures should be used when processing these samples to avoid exposure to diseases like hepatitis.

6. Interpretation of Bioassay Results

6.1 General Considerations

The preceding section describes a number of different bioassay measurements that can be made. Each of them may be used to detect the internal deposition of specific radionuclides, and to provide information that can be used in the estimation of the magnitude and distribution of the deposition and the cumulative organ doses resulting from this deposition. Regardless of the method chosen, interpretation of the results must be based on some knowledge of the patterns of organ deposition and retention for the specific form of the radionuclide involved. Interpretations can have differing levels of complexity depending on the uses to be made of the results.

The least complex of these interpretations can occur in a well controlled work situation with small amounts of air or surface contamination where the bioassay results are negative or low enough to indicate that good exposure-control practices have been followed. If the detection limit for the bioassay method used is sensitive enough to detect a small internal deposition, no further interpretation may be required.

When a positive bioassay result is obtained, it is important to know what constitutes a significant internal deposition to determine what additional action should be considered. Existing radiation-protection dose-limiting recommendations are based on generalized models for the deposition and retention of each radionuclide type and for the resulting absorbed doses to different tissues. As noted earlier, these models can be used to establish action points to trigger further investigation and action if necessary (see Section 7).

Routinely, positive bioassay results are first interpreted using generalized models. If the estimated internal deposition is lower than a predetermined action level, no further action may be required. In the case of an apparent high internal deposition, individual followup must be made to estimate the resulting dose for the individual involved because the pattern of deposition and retention may be substantially different from the generalized model used to determine permissible working limits and investigation levels. If therapeutic procedures are to be used to accelerate the removal of a radionuclide from the person, individualized sampling and interpretation methods will be required to determine the associated changes in estimated dose equivalent values for different body organs.

The interpretation of bioassay results can be made with respect to a number of different endpoints depending on the radionuclide involved and the bioassay methods used. The major endpoints include:

a) Body burden of radionuclide at the time of bioassay measurement.
b) Body burden of radionuclide immediately after intake.
c) Intake that resulted in observed body burden.
d) Distribution of radionuclide among organs at the time of bioassay measurement.
e) Projected patterns of radionuclide retention in organs of interest.
f) Cumulative dose equivalents that will be received by individual organs during control intervals such as:
   1) Between exposure and bioassay;
   2) First quarter of year after exposure;
   3) First year after exposure; and
   4) 50 years after exposure.[4]

Items that must be addressed in the interpretation of bioassay results are given in Table 6.1.

6.2 Examples of Bioassay Interpretations

6.2.1 *In-vivo Measurements of* ^{134}*Cs*

To illustrate the general strategy that might be employed in the interpretation of bioassay results, one radionuclide will be discussed for which both *in-vivo* measurements and *in-vitro* analyses can be made and interpreted with relative ease. Although the example will be based on a specific radionuclide for clarity of presentation, the underlying principles apply to a number of bioassay interpretations. The radionuclide chosen for this example is ^{134}Cs because *in-vivo* measurements can be made with relative ease and because the patterns of distribution and excretion are uncomplicated and easy to understand. Cesium is known to be rapidly transferable (soluble in body fluids) in almost all of its pure chemical compounds. When this is the case, the route of exposure (inhalation, ingestion or via a wound) is not a critical

[4] The 50-year control period is the standard exposure interval recommended by ICRP Publications 26 (ICRP, 1977) and 30 (ICRP, 1979).

TABLE 6.1—*General checklist for interpretation of bioassay results*

1. Radionuclide species
2. Physical form of radionuclide
3. Chemical form of radionuclide
4. Route of exposure
5. Previous exposure history
6. Exposure time and period
7. Sampling time after exposure
8. Bioassay information available
9. Age and health status of subject
10. Appropriate metabolic model
11. Appropriate model to estimate intake, dose equivalent rate, or committed dose equivalent

determinant of the resulting tissue distribution, retention and dose equivalent estimate. Cesium, a chemical homologue of potassium, is distributed throughout the soft tissue mass of the body. This metabolic pattern and the photon emissions from ^{134}Cs make several bioassay procedures feasible and useful.

Although the preceding discussion applies to transferable compounds of ^{134}Cs, it should be recognized that some accidental exposure situations might involve ^{134}Cs incorporated into a poorly transferable form such as a mixture of materials fused together in a high temperature reaction. The following example will show how the bioassay measurements can be used to ascertain the probable exposure form and its influence on the subsequent identification of tissues at risk and dose calculations.

As the result of a routine whole-body count, a person is determined to have a body burden of 1.7 μCi (6.3 × 10^{4}Bq)^{134}Cs. The possible interpretation of this result depends on whether the ^{134}Cs is in a transferable or nontransferable form in the body. Repetitive whole-body counts, made with various portions of the body shielded, indicate a widespread distribution throughout the body characteristic of the exposure to a transferable form of ^{134}Cs.

When did the exposure occur and what was the initial body burden at that time? A review of the work records does not indicate any unusual occurrences. The worker's last whole-body count, 100 days earlier, showed no detectable body burden of ^{134}Cs. In the absence of an obvious exposure date or unusual items in the individual's work history, the conservative assumption will be made here that the exposure occurred immediately after the last whole-body count.

From ICRP Publication 30 (ICRP, 1979), it is noted that the generalized, *biological*, whole-body retention function, $R_b(t)$, for non-

6.2 BIOASSAY INTERPRETATIONS / 35

radioactive cesium in a transferable form can be described with the following type of function:

$$R_b(t) = \frac{BB(t)_b}{BB(0)} = a_1 e^{-\lambda_{b_1} t} + a_2 e^{-\lambda_{b_2} t}, \quad (6.1)$$

where $BB(t)_b$ is the body burden at t days after exposure, $BB(O)$ is the initial body burden, $\lambda_{b_i} = 0.693/T_{b_i}$, and T_{b_i} is the biological half-life for the ith compartment. With the parameter values listed for cesium in Publication 30 ($a_1 = 0.1$, $a_2 = 0.9$, $T_{b_1} = 2$ d, and $T_{b_2} = 110$ d), this equation becomes:

$$R_b(t) = 0.1 e^{-0.35t} + 0.9 e^{-0.0063t}. \quad (6.2)$$

To use this equation for the whole-body retention of ^{134}Cs, the biological retention half-lives must be converted to effective half-lives. An *effective* half-life, T_e, is given by the relation:

$$T_e = \frac{T_b T_p}{T_b + T_p},$$

where T_p is the physical half-life. Because the physical half-life of ^{134}Cs is 748d, the effective half-lives for the two retention compartments for ^{134}Cs are:

$T_{e_1} = (2)(748)/(2 + 748) = 2$ d, and $T_{e_2} = (110)(748)/(110 + 748) = 96$d. With
$\lambda_{e_i} = 0.693/T_{e_i}$, the effective whole-body retention, $R_e(t)$, is

$$R_e(t) = \frac{BB(t)_e}{BB(O)} = a_1 e^{-\lambda_{e_1} t} + a_2 e^{-\lambda_{e_2} t}. \quad (6.3)$$

Using the appropriate values for ^{134}Cs, the generalized, *effective*, whole-body retention function is:

$$R_e = 0.1 e^{-0.35t} + 0.9 e^{-0.0072t} \quad (6.4)$$

With this retention function, a measured body burden of 1.7 μCi (6.3 × 10^4 Bq) at 100 d would correspond to an initial body burden of 3.9 μCi (1.4 × 10^5 Bq). This is shown graphically as curve A in Fig. 6.1.

Assume a second whole-body count was taken 100 d later resulting in 1.0 μCi (3.7 × 10^4 Bq) and denoted by point 2 in Fig. 6.1. This value would project back on curve B to an initial body burden of 4.7 μCi (1.7 × 10^5 Bq). Likewise, a measurement of 0.59 μCi (2.2 × 10^4 Bq) at 300 d, point 3, would be projected back on curve C to an initial body burden of 5.7 μCi (2.1 × 10^5 Bq).

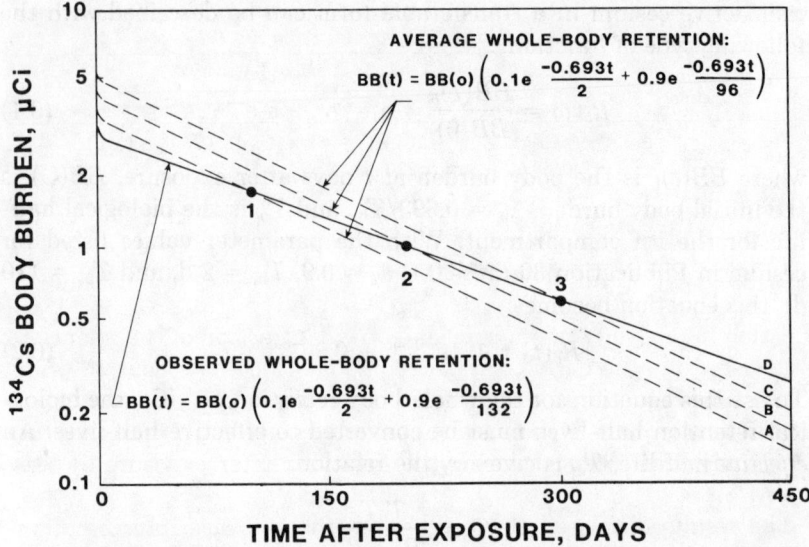

Fig. 6.1 Illustration of the determination of an initial ^{134}Cs body burden from three single whole-body measurements (points 1, 2 and 3 with curves A, B and C, respectively) and from three consecutive measurements (points 1, 2 and 3 with curve D). The equation for the generalized effective, whole-body retention function is used for the single point estimates, while the observed retention for the person is determined from the three consecutive measurements which include radioactive decay. The observed retention includes the first term of the uptake retention because the first measurement (point 1) is obtained long after the period during which the first term is significant (neglecting radioactive decay during this short period). (S.I. conversion: 1 μCi = 3.7 × 10^4 Bq.)

From Fig. 6.1, it is obvious that the actual retention curves at points 1, 2, and 3 are not adequately represented by a retention function having an effective half-life for the long-term component of 96 days. A single-component exponential function fitted to these three datum points indicates that the effective half-life in this person is actually 132 days, not 96 days.

Because all of these whole-body counts were made at a relatively long time after the exposure took place, no information is obtained about the first, rapidly cleared component. For ^{134}Cs in a transferable form, extrapolating the long-term retention component back to the estimated time of exposure will provide an estimate of the initial body burden that is about 90% of what the actual initial body burden may have been. For completeness, the estimate from the literature for the first component (half life = 2 d) can be added to the resulting retention equation (as was done in Fig. 6.1) to represent the entire retention equation as well as possible. Whether this last adjustment is important to the final interpretation depends on the complexity of the retention

TABLE 6.2—*Comparison of the Integrated Retained ^{134}Cs Activity Using Four Different Whole-Body Retention Estimates From Fig. 6.1*

Retention Function	Estimated BB (0)	Cumulated Activity at		
		90d	1 y	50 y
	(μCi)[a]	(μCi-d)	(μCi-d)	(μCi-d)
A	3.9	230	450	490
B	4.7	280	550	590
C	5.7	340	660	710
D	3.2	210	470	540

[a] S.I. conversion: 1 μCi = 3.7 × 10^4 Bq.

function for a given radionuclide and on what fraction of the retention is described by the early clearance component(s) that occurred prior to the first *in-vivo* measurement. In the present case, an estimate of the first component is included for completeness, but it increases the dose equivalent received during the first year after exposure by less than 1%.

The results discussed above illustrate the stepwise interpretation that is required for whole-body measurements. A properly performed whole-body count provides information on the body burden existing at that time. Preliminary estimates of the initial body burden and its subsequent retention can be made by using this value and a generalized retention function. More refined retention estimates require multiple whole-body counts at appropriate intervals.

To obtain some indication of the impact of the four different retention estimates associated with curves A, B, C, and D in Fig. 6.1, each one has been integrated to determine the cumulated values of μCi-days (3.7 × 10^4 Bq-days) and shown in Table 6.2. The dose equivalent received by the subject is directly proportional to this time integral of the ^{134}Cs retention function. The results in Table 6.2 show that the actual integrated retention (Function D) at 50 y is 10 percent higher than Function A but is 7 and 23 percent lower than Functions B and C, respectively. Whether these differences between generalized and individualized estimates are important must be determined on a case-by-case basis, considering factors such as the radionuclide, its effective half-life in the body, the exposure level involved, and the uncertainties inherent in the bioassay and modeling procedures used.

6.2.2 *In-vitro Measurements of ^{134}Cs*

In a routine bioassay program, it is possible that the first indication of an exposure may come from analysis of a routinely collected urine or feces sample. Interpretation of a positive value requires knowledge of the excretory rate and pattern as a function of time after exposure. A generalized expression for the excretory rate is the negative[5] derivative of the generalized, biological, whole-body retention function (Eq. 6.1). Thus, the biological rate of excretion (the fraction of the initial body burden excreted per day, ignoring radioactive decay) is:

$$E_b(t) = \frac{dR_b(t)}{dt} = a_1\lambda_{b_1}e^{-\lambda_{b_1}t} + a_2\lambda_{b_2}e^{-\lambda_{b_2}t}. \tag{6.5}$$

Multiplying Eq. (6.5) by $e^{-\lambda_p t}$ to allow for the radioactive decay of the ^{134}Cs, the *effective* rate of excretion, $E_e(t)$, including radioactive decay is:

$$E_e(t) = E_b(t)e^{-\lambda_p t} = a_1\lambda_{b_1}e^{-\lambda_{e_1}t} + a_2\lambda_{b_2}e^{-\lambda_{e_2}t}, \tag{6.6}$$

where $\lambda_{e_i} = \lambda_{b_i} + \lambda_p$.

For ^{134}Cs with $T_p = 748$ d and the values of a_1, a_2, λ_{b_1} and λ_{b_2} used in Eq. (6.2), this function is:

$$E_e(t) = 0.035e^{-0.35t} + 0.0057e^{-0.0072t}. \tag{6.7}$$

The resulting expression for total excretion must then be pro-rated as appropriate among the different excretory pathways. This information should be derived from the scientific literature or estimated from similar radionuclides and forms for which the values are available. In almost all cases, only excretion via urine and feces needs to be considered.

For an internal deposition of ^{134}Cs in a transferable form, about 0.8 of the total excretion is via the urine and about 0.2 is via the feces beyond the first 3 to 4 days after exposure (ICRP, 1968). With this proration, the generalized fractional excretory rates are:

$$U_e(t) = 0.8E_e(t) = 0.028^{-0.35t} + 0.0046^{-0.0072t} \tag{6.8}$$

and

$$F_e(t) = 0.2E_e(t) = 0.0070^{-0.35t} + 0.0011^{-0.0072t} \tag{6.9}$$

where $U_e(t)$ and $F_e(t)$ are expressed as fractions of the initial body burden of ^{134}Cs excreted per day in the urine and feces, respectively.

To obtain an expression for total excretion rate as a fraction of the body burden *at the time of excretion*, the effective fractional excretion-rate function, $E_e(t)$, must be divided by the effective whole-body retention function, $R_e(t)$. For cesium, the first components of $E_e(t)$ and $R_e(t)$ become unimportant beyond two weeks after exposure. Thus, at later times, the total daily excretion rate as a fraction of the body

---

[5] Since the retention is decreasing, the change in retention with time is *negative*, whereas the excretion rate is positive (greater than zero).

burden existing at that time is:

$$\frac{E_e(t)}{R_e(t)} = \frac{a_2 \lambda_{b_2} e^{-\lambda_{e_2} t}}{a_2 e^{-\lambda_{e_2} t}} = \lambda_{b_2}. \quad (6.10)$$

The daily urine and feces excretion rates expressed as fractions of the body burden at the time of excretion can be determined for cesium by using the prorations between urine and feces used for Eq. (6.8) and (6.9):

$$\frac{U_e(t)}{R_e(t)} = 0.8 \frac{E_e(t)}{R_e(t)} = 0.8 \lambda_{b_2} = (0.8)(0.0063) = 0.0050 \quad (6.11)$$

$$\frac{F_e(t)}{R_e(t)} = 0.2 \frac{E_e(t)}{R_e(t)} = 0.2 \lambda_{b_2} = (0.2)(0.0063) = 0.0013 \quad (6.12)$$

Let A_U or A_F = activity of a radionuclide found in a daily urine or feces sample, respectively.
Then:

$$\frac{U_e(t)}{R_e(t)} = \frac{\dfrac{A_U}{BB(O)}}{\dfrac{BB(t)}{BB(O)}} = \frac{A_U}{BB(t)} = 0.8\lambda_{b_2} \quad (6.13)$$

and

$$BB(t) = \frac{1.25 A_U}{\lambda_{b_2}} \quad (6.14)$$

Likewise,

$$\frac{F_e(t)}{R_e(t)} = \frac{\dfrac{A_F}{BB(O)}}{\dfrac{BB(t)}{BB(O)}} = \frac{A_F}{BB(t)} = 0.2\lambda_{b_2} \quad (6.15)$$

and

$$BB(t) = \frac{5 A_F}{\lambda_{b_2}} \quad (6.16)$$

Using $\lambda_{b_2} = 0.0063$, Eq. (6.14) and (6.16)

$$BB(t) = 198 \, A_U \quad (6.17)$$

and

$$BB(t) = 794 \, A_F \quad (6.18)$$

Continuing with the example started above, assume that instead of a positive whole-body count, a 24-hr urine sample from a worker was determined to contain 0.0065 µCi (2.4 × 10² Bq) ^{134}Cs. From Eq. (6.17), it would be estimated that the existing body burden at the time of excretion was 1.3 µCi (4.8 × 10⁴ Bq). Additional 24-hr urine collections made on the next two days confirm the presence of ^{134}Cs in the urine and yield the estimates of existing body burden given in Table 6.3.

Assume further that the time of exposure is unknown but that the last urine sample, collected 100 days ago, did not contain detectable ^{134}Cs. Again, the conservative assumption is that the exposure occurred immediately after the last urine sample had been collected. The sample collection times would then be 100 to 102 days after exposure. With the use of these values and Eq. (6.4), three estimates of the initial body burden are given in Table 6.3. These results show that urine sampling can be used to estimate the existing body burden, but the variability among repetitive samples is usually much higher than that of repetitive *in-vivo* measurements. Although the urine samples represent incremental or integrated samples over the collection period, they may be properly approximated as instantaneous excretion rates because Eq. (6.10) does not vary significantly during the sampling period.

Follow-up urine collections, made at monthly intervals, result in the ^{134}Cs values given in Table 6.4 and plotted in Fig. 6.2. Each of these values has also been used to estimate the daily body burden using the generalized urinary fractional excretion rate of 0.0050 of the existing body burden per day. However, a line fitted to these urinary data by a non-linear least squares method indicates that the urinary excretion is decreasing with an effective half-life of 126 d, not 96. In this case, the proper factor to calculate an individualized value of the daily body burden from a daily urine sample is not 0.0050, but is obtained from Eqs. (6.6), (6.10), and (6.11) as follows:

$$E_e(t) = 0.035 e^{-0.35t} + 0.0041 e^{-0.0055t} \tag{6.19}$$

$$\frac{E_e(t)}{R_e(t)} = 0.0046 \tag{6.20}$$

$$\frac{U_e(t)}{R_e(t)} = (0.8)(0.0046) = 0.0037. \tag{6.21}$$

Thus, by taking multiple urine samples over an extended period, it is possible to determine by least squares fit the appropriate excretion/retention half-life to use for this individual. The daily body burden

6.2 BIOASSAY INTERPRETATIONS / 41

TABLE 6.3—*Estimates of the Current and Initial Body Burdens of ^{134}Cs from Repetitive 24-Hour Urine Samples*

Sample Day	^{134}Cs 24-h Urine	Time After Exposure	Estimated Values[a]	
			Existing Body Burden	Initial Body Burden
	(μCi)[b]	(d)	(μCi)	(μCi)
1	0.0065	100	1.3	3.0
2	0.0055	101	1.1	2.5
3	0.0070	102	1.4	3.2
			$\bar{x} = 1.3$	2.9

[a] Based on Eqs. (6.4) and (6.17).
[b] S.I. conversion: 1 μCi = 3.7 × 10^4 Bq.

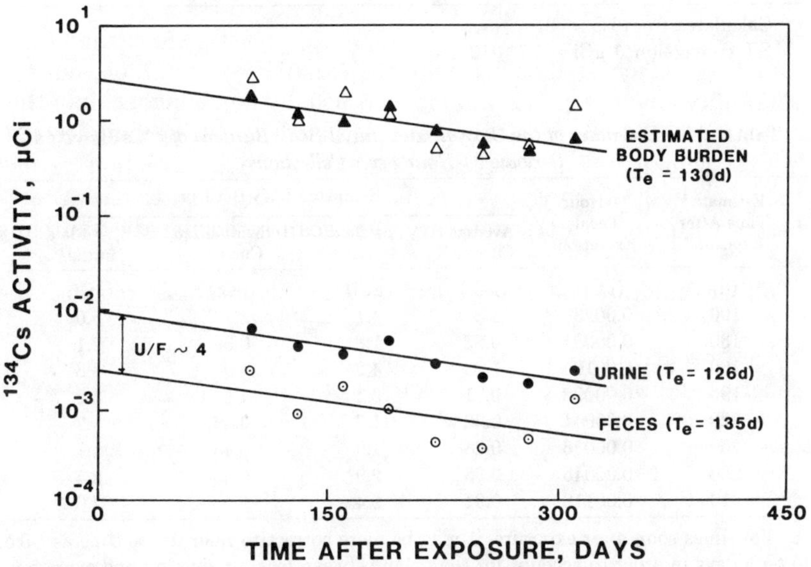

Fig. 6.2 Illustration of the relationship between ^{134}Cs urine and feces excretion and the ^{134}Cs whole-body burden estimated from the urine and feces excretion as a function of time. Each point in the urine and feces data represents the activity excreted per day on a given day while the body burden represents the activity in the person on the same day. (S.I. conversion: 1 μCi = 3.7 × 10^4 Bq.)

estimated with the generalized (T_e = 96 d) and individualized (T_e = 126 d) urinary excretion half-lives are given in Table 6.4.

A similar approach can be used with feces samples. In this example, the observed ^{134}Cs activity in feces samples collected on the same schedule as the urine samples are also plotted in Fig. 6.2 and given in Table 6.5. From Eq. (6.12), the generalized factor for estimating the

TABLE 6.4—*Estimates of the Current and Initial Body Burdens of ^{134}Cs Based on Periodic 24-Hour Urine Collections*

Estimated Time After Exposure	24-hour Urinary Excretion	Estimated Body Burdens			
		Average Current	$T_e = 96$ d Initial[a]	Individualized Current	$T_e = 126$ d Initial[a]
(d)	(μCi)[b]	(μCi)	(μCi)	(μCi)	(μCi)
100	0.0065	1.3	3.2	1.8	3.5
130	0.0041	0.82	2.3	1.1	2.5
160	0.0035	0.70	2.5	0.94	2.5
190	0.0048	0.96	4.2	1.3	4.1
220	0.0028	0.56	3.0	0.76	2.8
250	0.0020	0.40	2.7	0.54	2.4
280	0.0017	0.34	2.8	0.46	2.4
310	0.0023	0.46	4.8	0.62	3.8

[a] Calculated from Eq. (6.3).
[b] S.I. conversion: 1 μCi = 3.7 × 10^4 Bq.

TABLE 6.5—*Estimates of the Current and Initial Body Burdens of ^{134}Cs Based on Periodic 24-Hour Feces Collections*

Estimated Time After Exposure[a]	24-Hour Fecal Excretion	Estimated Body Burdens			
		Average Current	$T_e = 96$ d Initial[b]	Individualized Current	$T_e = 135$ d Initial[b]
(d)	(μCi)[c]	(μCi)	(μCi)	(μCi)	(μCi)
100	0.0023	1.8	4.1	2.7	5.0
130	0.00081	0.62	1.8	0.96	2.1
160	0.0016	1.2	4.2	1.9	4.8
190	0.00092	0.71	3.1	1.1	3.2
220	0.00041	0.32	1.7	0.49	1.7
250	0.00036	0.28	1.9	0.43	1.7
280	0.00045	0.35	2.9	0.53	2.5
310	0.00118	0.91	9.4	1.4	7.6

[a] For times soon after exposure, it may be more correct to refer to the time as t-0.5 to t-1.5 days in order to account for the delay between fecal production and excretion, depending on the individual's excretion habits.
[b] Calculated from Eq. (6.3).
[c] S.I. conversion: 1 μCi = 3.7 × 10^4 Bq.

existing body burden from a daily feces sample based on a 96-d effective retention half-life is 0.0013. Estimated daily body burden values derived with this factor are given in Table 6.5. However, as was the case with the urine samples, a least squares line fitted to the observed samples indicates an excretion/retention effective half-life of 135 d, not 96 d.

With an effective half-life of 135 d, the estimated equations for $E(t)$,

$\dfrac{E_e(t)}{R_e(t)}$, and $\dfrac{F_e(t)}{R_e(t)}$ becomes:

$$E_e(t) = 0.035e^{-0.35t} + 0.0038e^{-0.0051t} \qquad (6.22)$$

$$\dfrac{E_e(t)}{R_e(t)} = 0.0042 \qquad (6.23)$$

$$\dfrac{F_e(t)}{R_e(t)} = (0.2)(0.0042) = 0.00084 \qquad (6.24)$$

The use of the feces excretion factor given in Eq. (6.24) yields the estimates of current body burden of ^{134}Cs listed in Table 6.5 in the column headed $T_e = 135$ d.

All body burden values estimated from the measurements of the daily urine and feces samples and the observed excretion half-lives are plotted in Fig. 6.2. It can be seen that, in this case, both the urine and feces samples have provided equally useful estimates of the daily body burden. The best estimate of the whole-body retention of ^{134}Cs from the excreta samples *of this person* can be obtained by fitting a least squares curve to all the available values. This single exponential function has a slope corresponding to a 130-d effective retention half life and an initial burden at 100 d before the first sample of 2.7 µCi (10^5 Bq).

The choice of whether one should use the results from urine samples or feces samples or both to estimate body burdens is heavily dependent on the circumstances involved with the particular exposure being evaluated including intake route. The metabolic characteristics of the exposure material must be considered as well as the number and quality of various kinds of samples that are available. For materials that are readily soluble in body fluids, the use of urinary excretion values may be more reliable for assessing systemic burden whereas for very insoluble materials, the use of data from the fecal samples may be best for assessing lung clearance after inhalation. Because there is no one answer for all possible exposure situations, careful attention should be given to choosing from the various analyses that are available.

The foregoing portion of the example with ^{134}Cs is based on the assumption that the ^{134}Cs had been deposited and instantaneously taken up into the circulation in a transferable form. Although this is a reasonable first assumption, checks should be made to determine whether it is correct. If the ^{134}Cs were inhaled in a poorly transferable form, practically all of the body burden remaining after clearance of the nasopharyngeal and tracheobronchial regions of the respiratory tract (3 to 4 days post-exposure) would be in the pulmonary region where it might remain with a very long (e.g., 500 d) biological half-life. Two main characteristics of the *in-vivo* measurements can be used to ascertain that ^{134}Cs is in a poorly transferable form:

a) The detectable ^{134}Cs would be primarily associated with the thoracic region instead of being distributed relatively widely throughout the body.
b) The observed retention half life would be greater than ~200 days.

The excretion data could also be used for this purpose. If only urinary excretion data are available, the judgment would have to be made based on an observed urinary excretion half-life of more than ~200 days, but this would take a long time to determine. (The situation is actually more complex because the lung-transfer mechanisms will be superimposed on the whole-body metabolism.) An indication could be obtained more quickly by collecting both urine and feces samples at the same time. As illustrated in Fig. 4, the amount of transferable ^{134}Cs excreted in the urine usually exceeds that excreted in the feces by a factor of 2 to 10 (*i.e.*, $U/F = 2$ to 10). For poorly transferable ^{134}Cs, where the main reservoir is the pulmonary region, excretion via the feces exceeds that of the urine and $U/F < 1$. This information would provide a quick indication that the ^{134}Cs was not behaving as expected for a transferable (soluble) compound of ^{134}Cs and that special precautions would have to be taken when calculating the dose commitment resulting from such an exposure.

As a third approach, it may be feasible in some cases to obtain *in-vitro* dissolution data on the material that will aid in predicting its *in-vivo* retention (Kanapilly *et al.*, 1973, 1974; Moss and Kanapilly, 1980).

6.3 Other Radionuclides

The foregoing example represents one of the easiest bioassay situations that might be encountered because it involves a radionuclide having the following characteristics:

1) Ease of detectability in both excreta and *in vivo*.
2) Uncomplicated metabolic pattern.
3) Clear distinction between transferable and non-transferable forms.

6.3 OTHER RADIONUCLIDES / 45

Most of the other radionuclides are more difficult to work with because they lack one or more of the above attributes. As pointed out previously in Section 5.3, bioassay of ^{3}H is easily done with urine analyses but cannot be done by *in-vivo* measurements because of the lack of penetrating radiations. The radionuclide ^{239}Pu is representative of a number of actinide radionuclides for which the interpretation of bioassay measurements is difficult, regardless of the method used. *In-vivo* measurements can be influenced substantially by the actual distribution of ^{239}Pu in the lung and in the rest of the body as well as by the shape and thickness of the thorax being measured. Also, even under ideal situations, *in-vivo* methods are not sensitive enough to measure all quantities of ^{239}Pu that are of interest for bioassay purposes. Interpretation of results from urine and feces sample analyses is highly dependent on the form of ^{239}Pu in the body and the main reservoirs of ^{239}Pu at the time of sampling.

One of the most complicated bioassay situations for which information has been published has been the bioassay of uranium miners for their exposure to radon and its decay products. In this situation, one must deal with the whole decay chain of radon progeny. Various approaches to bioassay for ^{210}Pb or ^{210}Po have been used, followed by back calculations to the original exposure. Approaches used have included external skeleton (skull) counting of ^{210}Pb, and radiochemical analyses of urine, feces, whiskers, blood, and autopsy specimens. Reports such as Black *et al.* (1968), Gotchy and Schiager (1969), Blanchard and Moore (1971), Schiager and Savignac (1972), Cohen *et al.* (1973), Blanchard *et al.* (1973), and Pomroy and Measures (1979) provide an interesting perspective on this difficult problem.

The most important point to be gained from this section is that the establishment of bioassay procedures for a radionuclide and their subsequent interpretation are dependent upon the physical, chemical and biological characteristics of that radionuclide. Each radionuclide should be considered separately, but the details relevant to individual radionuclides are beyond the scope of this report. The reader is encouraged to use existing bioassay formulations as well as the current scientific literature (*e.g.*, Broga *et al.*, 1986; Leggett, 1986; Cattarin and Doretti, 1986) concerning the metabolic, measurement and other aspects of the radionuclide involved.

7. Action Points and Action Based on Bioassay Results

Appropriate action, based on bioassay results, is dependent on the underlying purpose of the measurement. The following material describes these purposes and actions.

7.1 Preparatory Evaluation

Where bioassay is used to screen personnel prior to a job assignment, the presence of radioactive material in excess of natural background or fallout levels, as detected by routine bioassay procedures, should trigger an investigation. Information regarding the location and quantity of radioactive material in the body should be sought, and conservative predictions as to future retention in the body should be made. This information can usually be derived from a review of the worker's previous exposure history, including previous bioassay results, and from subsequent bioassay measurements as necessary. Findings should be compared with the criteria of dose-limiting recommendations, and a decision should be made to approve the job assignment if the findings are acceptable, or, otherwise, to impose a delay or a restriction.

7.2 Exposure Control

When, in the course of the worker's employment, bioassay measurements are being made routinely, it is essential to ensure that the results are carefully reviewed by qualified personnel and that appropriate action is taken if the results are considered high. Action should be based on the organ or body burden, the committed dose equivalent or, in the case of uranium, on possible chemical damage to the kidneys as indicated by the result.

Working conditions often involve daily exposure to low levels of airborne radioactive material with little variability in the concentration in air. Under these exposure conditions, the body burdens of some

7.2 EXPOSURE CONTROL / 47

radionuclides may gradually build up until the rate of elimination is virtually equal to the rate of accumulation. When bioassay results indicate that the body burden is continuing to rise, action should be taken to assure that additional buildup will not interfere with the worker's career. For exposure to non-transferable materials that can be measured by *in-vivo* techniques, work restrictions are normally imposed on the basis of these measurements, and, hence, urinalysis results are primarily used to indicate the immediate need for *in-vivo* measurements.

Examples of appropriate corrective actions are shown in Table 7.1. These actions are based on arbitrarily chosen fractions of an unspecified basic protection criterion related to intake or organ burden levels. Table 7.1 specifies that no action is required if the bioassay result indicates that $1/10$ to $1/4$ of the protection criterion will not be exceeded. It is recognized that, at present, the quantity of activity associated with $1/10$ of the protection criterion for many radionuclides cannot be detected, or can be detected only through employment of prohibitively expensive laboratory procedures. For such nuclides, the fraction of $1/10$ must be increased to conform to the minimum detectable level, and reliance for protection must be placed primarily on other measurements such as air sampling.

If the operations being monitored are generally well controlled, and levels of air or surface contamination are low, the results of bioassay on the workers involved can be taken as indices of good practices. If a bioassay result is then obtained that is significantly in excess of these index values, it is an indication of some abnormal exposure or condition that should be investigated. This is a useful approach where a number of workers, such as 10 or more, work under similar conditions. All these workers would be expected to show about the same bioassay results, and any individuals consistently showing excessive values should be the subject of an investigation of their work practices. If work conditions remain generally constant over a period of several years, action points can be developed based on the degree of departure from the bioassay norm.

For other operations that are not well controlled, bioassay results can still be related to values obtained in well-controlled areas. This approach is most useful when the exposure conditions are chronic; *i.e.*, a generally constant level of exposure persisting over a long period of time. However, even with intermittent exposure, as from occasional accidental releases, bioassay results may indicate a need for investigation even in the unlikely event that such releases are not detected by air or surface contamination monitoring. Even where all the exposed personnel show bioassay results well below action levels based

TABLE 7.1—*Examples of Corrective Actions Based on Bioassay Measurement Results*

Fraction of Protection Criterion	Interpretation	Actions
Below 1/10	Confinement and air sampling capabilities confirmed.	None required (although it is good practice to investigate a positive bioassay result when none were being observed).
Between 1/10 to 1/4	Confinement and/or air sampling capabilities marginal.[a]	1. Confirm result (repeat measurement). 2. Identify cause and initiate additional control measures 3. Determine whether other workers could have been exposed and perform bioassay measurements for them. 4. If exposure (indicated by excreta analysis) could have been to Class (Y) material that can be measured using *in-vivo* techniques, consider such measurements.[b]
Between 1/2 and 1	Confinement and/or air sampling capabilities unreliable[a]	1. Take the actions listed above. 2. If exposure (indicated by excreta analysis) could have been to Class (Y) material that can be measured using *in-vivo* techniques, assure that such measurements are performed.[b] 3. Determine why air samples were not representative and did not warn of excessive airborne radioactivity. Make necessary corrections. 4. Perform additional bioassay measurements as necessary to make a preliminary estimate of the critical organ burden; consider work assignment limitations that will permit natural elimination and/or radioactive decay and assure that the protection criterion is not exceeded. 5. If exposure could have been to Class (Y) material, continue operations only if it is virtually certain that the protection criterion will not be exceeded by any worker.[b]
Greater than 1	Containment and/or air sampling capabilities unacceptable.[a]	1. Take all actions listed above. 2. Continue operations only if it is virtually certain that the protection criterion will not be exceeded by any other worker, regardless of radionuclide classification. 3. Establish work assignment limitations as necessary for affected workers. 4. Perform individual metabolic studies for affected workers.

[a] Unless the result was anticipated and caused by conditions already corrected.
[b] Material classification per ICRP Publication 30 (ICRP, 1979).

on metabolic considerations and radiation dose, any departure from the norm should be viewed with suspicion and may warrant an investigation of the cause. In fact, this empirical approach is the essence of achieving results in the effort to maintain the as low as reasonably achievable level.

7.3 Diagnostic Evaluation of Bioassay Measurements

Diagnostic bioassay measurements are made to estimate the quantity and distribution of radionuclides in the body after determination that a large or unusual deposition has occurred. Actions to be based upon diagnostic results include:

a) Selection of subsequent measurement methods and frequencies;
b) Imposition or removal of work restrictions;
c) Referral to a physician;
d) Evaluation by the physician of the need to accelerate the elimination of the radionuclide;
e) Modification of the physiological model using data from the individual being studied; *i.e.*, determination of the individual's own retention and excretion rates.

Excellent examples of how a variety of bioassay procedures can be used advantageously to assess an accidental exposure case and guide treatment are described in the published reports related to the 1976 Hanford americium exposure incident (Palmer and Rieksts, 1983; Robinson *et al.*, 1983).

8. Perspective on Bioassay

The preceding information in this report has indicated what bioassay is, why it should be done, on whom it should be done, ways in which it can be done, how often it should be done, and how the results might be interpreted. A consistent theme throughout this report is the view that bioassay is a vitally important part of a complete radiation protection program because it is aimed at determining whether a worker's body contains any internally deposited radionuclides and, if so, how much. Information of this type is useful for controlling the amount of radiation received from internally deposited sources. This information can also be used for indirect monitoring of the exposure control procedures in use and, if necessary, to establish appropriate therapeutic procedures.

This report stresses the importance of bioassay, but it is also necessary to recognize the limitations of bioassay procedures. The most important parameter in radiation protection is the dose equivalent received by individual organs and local tissues within them. Existing bioassay techniques do not provide a means of measuring these doses directly.

Estimates of radionuclide intakes or body burdens and their associated dose equivalents are limited by errors in the measurements made, uncertainties in the models used to interpret the results, and variability among subjects in radionuclide metabolism and dosimetry. Each of these topics can have a profound influence on the overall reliability of the resulting internal dose assessment.

8.1 Errors in Measurement

As with any analytical measurement, one must be concerned with both the accuracy and precision of the bioassay measurements made. It is particularly difficult to achieve high values of these when dealing with samples whose radionuclide levels are near the minimum detectable amount, MDA, for the method being used. Also, for samples with a radionuclide content below the MDA, a certain amount of internal radionuclide deposition could be undetected, and this would be another source of error.

The material presented in Section 5 discussed a number of error sources for various types of bioassay measurements. For *in-vivo* measurements, these included the type of detection system used for the type and energies of radiation emitted, its calibration, and the consistency and magnitude of background radiation detected by the system. Some other sources of error for a particular subject are his or her body geometry as compared to the calibration geometry, the magnitude of the internal deposition, and its distribution within the subject's body.

For *in-vitro* analyses of urine or feces, errors associated with sample collection include the representative nature of the sample analyzed (*e.g.*, was it a true 24-hr urine collection?), possibilities of sample cross contamination or losses, and errors resulting from other sample handling and processing procedures. The magnitude of the analytical recovery and its stability are also important factors in assessing possible errors in measurement for analyses of biological specimens such as urine and feces.

8.2 Uncertainties in Internal Dose Assessment

Having made the appropriate analytical measurements and assessed their associated errors, one must use various deposition, retention, and dosimetric models to work back to the existing body or organ burden or to the intake(s) that resulted in these burdens. Different models have different degrees of appropriateness for a particular bioassay application and the user should be aware of them and the uncertainties associated with various models. These uncertainties relate to both the adequacy with which a given model portrays the metabolism and internal dosimetry of a given radionuclide exposure for a standard man and also the variability seen among people that all received the same exposure.

When selecting bioassay and dosimetric models, one should determine what models are currently available, the data bases on which they were constructed, and the radionuclides, forms, and exposure conditions for which they are applicable. Typical questions that should be addressed include: Was bioassay one of the purposes for which the model was originally designed? Is the model specific for the situation being assessed? Is it based on adequate data from exposed human or laboratory animal populations? Is it derived by extrapolation from results for other radionuclides or exposure scenarios? If it is an extrapolation, is it reasonable?

General metabolic and dosimetric models have been promulgated by the ICRP in its Publication 30 (ICRP, 1979) and conditionally affirmed by the NCRP (NCRP, 1985) to establish derived air concentrations and annual limits on occupational intakes for radiation protection planning purposes. To assure adequate protection of all workers, these models involve many generalizations and conservative assumptions. ICRP Publication 30 points out that "The models used in this report have been chosen, often conservatively, to derive values of ALI or DAC to ensure the protection of workers. The data provided herein should therefore not be used indiscriminately out of context, e.g., to estimate the risk of cancer in individual cases." The NCRP affirmed this is their Report No. 84 (NCRP, 1985). Examples of modifications of general models for individuals are contained in Leggett (1986) for cesium and Broga et al. (1986) for iodine. Thind (1986) points out the importance of particle size assumptions in the use of ALI and DAC values.

For many radionuclide exposure situations, definitive bioassay models are not currently available. In such a situation, one might use the generalized metabolic and dosimetric information given in ICRP Publications 10, 10A, and 30 to interpret their results. The user should be aware that the use of generalized metabolic and dosimetric parameters of this type for internal dose assessment involves uncertainty, the magnitude of which may be quite large, depending on the conditions being assessed.

Such an approach may be adequate for routine monitoring purposes. However, it is desirable to reduce this uncertainty as much as possible when evaluating an individual potential exposure case. Obviously one needs to have a higher degree of accuracy when dealing with a suspected over-exposure to ensure that the assessment of internal committed dose and consideration of therapeutic removal procedures are adequate for the exposure conditions involved. Steps that might be taken to reduce this uncertainty might involve the use of more specific and detailed metabolic or dosimetric models for the particular exposure conditions. If better models are not available, it may be helpful to adjust certain parameters in the models being used to reflect the observations and measurements made in this particular case.

Variability among individuals in a population is also a major source of uncertainties. The generalized metabolic and dosimetric results given in ICRP Publication 30 are based on the use of anatomical and physiological data for Reference Man as detailed in ICRP Publication 23 (ICRP, 1975). However, it is obvious that an individual exposure situation may involve a person differing substantially from Reference Man in any of the parameters characterized. Some idea of the potential variability can be gained from the wealth of supporting information given in ICRP Publication 23. Further information on intersubject variability in anatomical and biochemical features has been published by Williams (1956) and in other sources.

The subject of variability in depositions of toxic substances in target organs was addressed by Cuddihy et al., 1979. The analyses reported were based both on results from human populations and studies in laboratory animals exposed under well-controlled experimental conditions. In inhalation exposure studies, about 2% of the animals exposed under identical conditions may receive more than three times the average organ doses of the whole group. For ingestion studies, 2% received more than two times the average doses. Analyses of human populations exposed by inhalation to environmentally dispersed materials also showed that a small percentage of the individuals in the population may receive up to five times the average organ doses received by the population. The changes in model parameters mentioned above would be for the purpose of improving the quality with which the models represented what was taking place in the subject being studied.

Another kind of uncertainty associated with the calculation and interpretation of bioassay results is that associated with the time at which the exposure occurred. If no indication of the possible exposure time is obtained from the worker or the associated work records, an exposure time must be estimated, and this can lead to different amounts of intake or committed dose being accounted for. Johnson (1985) provided an illustration showing that different amounts of intake or committed dose equivalent can be unaccounted for depending on the annual sampling frequency, the ratio of the length of the sampling interval to the radionuclide's effective half-life, and the assumption used for the time when exposure occurred.

The ICRP, in its Publication 26 (ICRP, 1977), recommends individual monitoring for persons working under conditions in which the dose equivalent or intake might exceed 3/10 of the annual limit. An investigation level is defined by them as the value of dose equivalent or intake above which the results are considered significantly important to justify further investigations. The derived investigation level for internal radionuclide deposition corresponding to the monitoring recommendation above is DIL = 0.3 ALI/f where ALI = annual limit or intake, and f = annual sampling frequency.

Assuming that the exposure time is unknown, the ICRP (1982) recommends that the intake be assumed to have occurred at the midpoint of the interval since the last bioassay sampling. Johnson (1985) compared the dosimetric consequences of using (a) the mid-point and (b) the beginning of the most recent bioassay interval for the time of exposure in the associated bioassay calculations. His analysis showed that the amount of committed dose equivalent or intake that might be unaccounted for by using the above definition of a derived investigation level is a function of the ratio of the length of the sampling interval to the effective half life of the radionuclide involved. When the mid-point of the interval is used, the amount of committed dose equivalent or intake that is unaccounted for continues to increase as the ratio of sampling interval length to the radionuclide's effective half-life increases. By choosing the more conservative approach of assigning the exposure to the beginning of the interval, the amount of committed dose equivalent or intake that is unaccounted for is limited to (0.3)(annual limit)/f.

8.3 Adequacy of Interpretation of Bioassay Results

As can be seen from the foregoing material, the collection, analysis, and interpretation of bioassay information is a complex subject best handled by individuals who are knowledgeable in this field. A good understanding not only of the mechanics of the bioassay process but also of the problems and pitfalls associated with the results obtained is needed.

APPENDIX A

Glossary

anatomic models: Stylized descriptions of the location, size, and shape of selected portions of the anatomy of man, woman, or child.

annual limit on intake (ALI): The activity of a radionuclide that taken alone would irradiate a person, represented by Reference Man, to the limit set for occupational exposure by recommending and regulating bodies.

bioassay procedure: A procedure used to determine the kind, quantity, location and/or retention of radionuclides in the body by direct (*in-vivo*) measurements or by *in-vitro* analysis of material excreted or removed from the body.

biokinetic model: A series of mathematical relationships formulated to describe the intake, uptake and retention of a radionuclide in various organs of the body and the subsequent excretion from the body by various pathways.

biological half-life (T_b): The time required for a biological system, such as a person, to eliminate by natural process, other than radioactive decay, one-half of the amount of a substance, such as a radionuclide, that has entered it.

body burden: The quantity of radioactive material contained in the individual's body at a particular point in time.

chelation therapy: Procedures used to remove an internally deposited radionuclide from a person's body through the administration of a metal chelating agent to enhance excretion of the radionuclide. The most common chelating agent in use today is a salt of diethylenetriaminepentaacetic acid, DTPA.

committed dose equivalent (H_{50}): The total dose equivalent averaged throughout any tissue in the 50 years after intake of a radionuclide into the body.

critical organ: The organ determining the maximum permissible concentrations in air or water. Theoretically it was the organ suffering the most damage or whose functions were most essential, but in practice it was usually the organ with the highest concentration of radionuclide.

derived air concentration (DAC): Equals the ALI of a radionuclide divided by the volume of air inhaled by Reference Man in a working year (*i.e.*, 2.4×10^3 m^3).The unit of DAC is Bq m^{-3}.

derived investigation level: A level associated with quantities other than dose equivalent, committed dose equivalent or intake which, by use of a defined model, is intended to reflect the investigation level in a particular situation in the workplace (see *investigation level*).

derived limits: Any of a number of limits such as "Derived Air Concentration" (DAC) calculated from the basic radiation protection limits by application of standard parameters for the pertinent biological system.

direct bioassay: The assessment of radioactive material deposited in the body by detection of radiations emitted by the material (see *in-vivo* methods).

dose equivalent (H): A quantity used for radiation protection purposes that expresses on a common scale for all radiations, the irradiation incurred by exposed persons. It is defined as the product of the absorbed dose and certain modifying factors. The unit of dose equivalent is currently the rem (see rem) or the sievert (see Sv).

effective half-life (T_e): The time required for the amount of a radionuclide deposited in a living organism to be diminished 50 percent as a result of the combined action of radioactive decay and biological elimination,

$$i.e., \ T_e = \frac{T_b \ T_p}{T_b + T_p}$$

exposure route: A pathway by which a radionuclide or other toxic material can enter the body. The main exposure routes are inhalation, ingestion, absorption through the skin, and entry through a cut or wound in the skin.

gray (Gy): The unit of absorbed dose in the SI (NCRP, 1985): 1Gy = 1 J kg^{-1} = 100 rad.

indirect bioassay: The assessment of radioactive material deposited in the body by detection of radioactivity in material excreted or removed from the body (see *in-vitro* methods).

insoluble material: A term loosely used to describe the relative degree of solubility of a material in body fluids. Recognizing that no material is absolutely insoluble, the terms low solubility or poorly soluble are preferable.

intake: The amount of radioactive material taken into the body by inhalation, absorption through the skin, injection, ingestion, or through wounds.

GLOSSARY / 57

internal emitter: A term used for a radionuclide deposited in the body.
investigation level: A value of dose equivalent or radionuclide intake above which the result is sufficiently important to justify further investigation.
***in-vitro* methods:** Detection of radiations emitted by radioactive materials excreted or removed from the body, using radiochemical and/or radioanalytical techniques.
***in-vivo* methods:** Detection of radiations emitted by radioactive materials deposited in the body, usually by whole-body (or critical organ) counting techniques.
maximum permissible concentration (MPC): The concentration of air (MPC_a) or water (MPC_w) that would lead at equilibrium to an amount of radionuclide in the critical organ that would just deliver the maximum permissible dose rate to that organ. These were calculated for either 40 hrs. per week or 168 hrs. per week intakes.
metabolic model: A mathematical description of the behavior in the metabolic processes of cells, tissues, organs and organisms. It is most frequently used to describe distribution among tissues and excretion.
minimum detectable amount (MDA): A term used to indicate the ability of a particular radioanalytical method to discern between the radioactivity associated with the radionuclide content of the sample of interest and background radiation with a particular degree of statistical confidence.
nuclide: A species of atom characterized by the constitution of its nucleus.
organ burden: The quantity of a radionuclide present in an organ at a particular point in time.
physical half-life (T_p): The time required for the level of radioactivity of a radionuclide to decrease by 50 percent due only to natural radioactive decay processes.
rad: A unit of absorbed dose: One rad is 0.01 joules absorbed per kilogram of any material. (Also defined as 100 ergs per gram.) Being replaced by the gray (NCRP, 1985).
radionuclide: A nuclide that is radioactive.
radionuclide exposure: The situation leading to intake of a radionuclide and/or the situation after a radionuclide has been deposited in an organ or tissue.
Reference Man: A person with the anatomical and physiological characteristics defined in the report of the ICRP Task Group on Reference Man (ICRP Publication 23).

rem: A unit of dose equivalent. The product of the absorbed dose in rad and modifying factors. Being replaced by the sievert (NCRP, 1985).

retention function: A mathematical expression for the fractional retention of a nuclide in an organ or body at any time following intake.

sievert (Sv): The measure of dose equivalent in the SI. It is the product of absorbed dose in grays and any modifying factors. It is equivalent to 100 rem.

uptake: Quantity of a radionuclide taken up by the systemic circulation, e.g., by injection into the blood, by absorption from compartments in the respiratory or gastrointestinal tracts, or by absorption through the skin or through wounds in the skin.

REFERENCES

ALEXANDER, R. E. (1974). *Applications of Bioassay for Uranium.* USAEC Report No. WASH-1251 (U.S. Government Printing Office, Washington, D.C.).

ANSI (1978). *American National Standard for Internal Dosimetry for Mixed Fission and Activation Products,* ANSI N343-1978 (American National Standards Institute, New York, New York).

ANSI 1983. *American National Standard For Dosimetry-Internal Dosimetry Programs for Tritium Exposure-Minimum Requirements,* ANSI N13.14-1983 (American National Standards Institute, New York, New York).

AUB, J. C., EVANS, R. D., HEMPELMANN, L. H., AND MARTLAND, H. S. (1952). "The late effects of internally deposited radioactive materials in man," Medicine **31,** 221.

BLACK, S. C., ARCHER, V. E. AND DIXON, W. C. (1968). "Correlation of radiation exposure and lead-210 in uranium miners," Health Phys. **14,** 81.

BLANCHARD, R. L. AND MOORE, J. B. (1971). "Body burden, distribution and internal dose of ^{210}Pb and ^{210}Po in a uranium miner population," Health Phys. **21,** 499.

BLANCHARD, R. L., KAUGHMAN, E. L. AND IDE, H. M. (1973). "Lead-210 concentration as a measure of uranium miner exposure," Health Phys. **25,** 129.

BODDY, K., ELLIOTT, A., ROBERTSON, I., MAHAFFY, M. E. AND HOLLOWAY, I. (1975). "A high sensitivity dual-detector shadow-shield whole-body counter with an invariant response for total body *in vivo* neutron activation analysis," Phys. Med. Biol. **20,** 296.

BOECKER, B. B. (1969). "The metabolism of ^{137}Cs inhaled as ^{137}CsCl by the beagle dog," Proc. Soc. Exptl. Biol. Med. **130,** 966.

BRADY, D. N. AND SWANBERG, F., JR. (1965). "The Hanford mobile whole-body counter," Health Phys. **11,** 1221.

BRODSKY, A. (1969). "Determination of facilities, equipment, and procedures required for various types of operations," page 664 in *Handbook of Radioactive Nuclides,* Y. Wang, Ed. (CRC Press, Boca Raton, Florida).

BRODSKY, A. (1980). "Resuspension factors and probabilities of intake of material in process (or is 10^{-6} a magic number in health physics)," Health Phys. **39,** 992.

BRODSKY, A. (1983). *Information for Establishing Bioassay Measurements and Evaluation of Tritium Exposure,* NUREG-0938 (U.S. Nuclear Regulatory Commission, Washington, D.C.).

BROGA, D. W., BERK, H. W. AND SHARPE, A. R., JR. (1986). "Efficacy of radioiodine urinalysis," Health Phys. **50,** 629.

BUISMAN, A. S. K. (1985). "On the frequency of whole-body counter monitoring," page 177 in *Assessment of Radioactive Contamination in Man 1984*, IAEA Publication No. STI/PUB/674 (International Atomic Energy Agency, Vienna).

CATTARIN, S. AND DORETTI, L. (1986). "A quick method for the assessment of internal contamination by ^{45}Ca," Health Phys. **50**, 523.

COHEN, N., JAAKKOLA, T. AND WRENN, M. E. (1973). "Lead-210 concentrations in the bone, blood and excreta of a former uranium miner," Health Phys. **24**, 601.

CUDDIHY, R. G., MCCLELLAN, R. O. AND GRIFFITH, W. C. (1979). "Variability in target organ deposition among individuals exposed to toxic substances." Toxicol. Appl. Pharm. **49**, 179.

CURTISS, L. F. AND DAVIS, F. J. (1943). "A counting method for the determination of small amounts of radium and of radon," J. Res. Nat. Bur. Stand. **31**, 181.

DANCER, G. H. C. AND FINNIGAN, T. (1975). "Interpretation of carbon-14 excretion data in terms of absorbed dose," page 83 in *Radiation Protection Measurement: Philosophy and Implementation*, Recht, P. and Lakey, J. R. A., Eds., Report EUR-5397e (Commission of the European Communities, Luxembourg, Belgium).

EAKINS, J. D. AND MORGAN, A. (1964). "The role of fecal analysis in a bioassay programme," page 231 in *Assessment of Radioactivity in Man*, Vol. 1, IAEA Publication No. STI/PUB/84 (International Atomic Energy Agency, Vienna).

EPSTEIN, R. J. AND JOHANSON, E. W. (1966). "Apparatus for monitoring ^{239}Pu in wounds," Health Phys. **12**, 29.

EVANS, R. D. (1935). "Apparatus for the determination of minute quantities of radium, radon and thoron in solids, liquids and gases," Rev. Sci. Instrum. **6**, 99.

EVANS, R. D. (1937). "Radium poisoning: II. The quantitative determination of the radium content and radium elimination rate of living persons," Am. J. Roentgenol. Radium Ther. **37**, 368.

FRIBERG, L. (1985). "The Rationale of Biological Monitoring of Chemicals with special Reference to Metals" Am. Ind. Hyg. Assoc. J. **46**, 633.

FROMHEIN, O., OHLENSCHLAGER, L. AND RAPP, W. (1976). "A new approach for improved short-time procedure for locating and measuring low-activity, low-energy transuranium deposits in wounds," page 223 in *Diagnosis and Treatment of Incorporated Radionuclides*, IAEA Publication No. STI/PUB/411 (International Atomic Energy Agency, Vienna).

GAUTIER, M. A., Ed. (1983). *Manual of Analytical Methods for Radiobioassay*, DOE Report No. LA-9763-M (National Technical Information Services, Springfield, Virgina).

GOTCHY, R. L. AND SCHIAGER, K. J. (1969). "Bioassay methods for estimating current exposures to short-lived radon progeny," Health Phys. **17**, 199.

GUILMETTE, R. A. (1986). "A low-energy photon detector for the radioassay of Pu and Am in biological samples, Health Phys. **51**, 797.

HALL, R. M., PODA, G. A., FLEMING, R. R. AND SMITH, J.A. (1978). "A mathematical model for estimation of plutonium in the human body from urine data influenced by DTPA therapy," Health Phys. **34,** 419.

HARLEY, J. H., ED. (1972). *HASL Procedures Manual*, USAEC Report HASL-300 (National Technical Information Service, Springfield, Virginia).

HEID, K. R. (1983). "A comparison of systemic burdens at autopsy to estimates based on health physics data for selected plutonium workers," Health Phys. **44,** Suppl. 1, 477.

HICKEY, J. L. S. AND CAMPBELL, S. D. (1968). "Mathematical model for determining low radium-226 body burdens from radon-222 in the breath," Health Phys. **14,** 119.

IAEA (1962). International Atomic Energy Agency, *Whole-Body Counting*, IAEA Publication No. STI/PUB/47 (International Atomic Energy Agency, Vienna).

IAEA (1970). International Atomic Energy Agency, *Directory of Whole-Body Radioactivity Monitors*, IAEA Publication No. STI/PUB/213 (International Atomic Energy Agency, Vienna).

IAEA (1972). International Atomic Energy Agency, *Assessment of Radioactive Contamination in Man*, IAEA Publication No. STI/PUB/290 (International Atomic Energy Agency, Vienna).

IAEA (1976). International Atomic Energy Agency, *Diagnosis and Treatment of Incorporated Radionuclides*, IAEA Publication No. STI/PUB/411 (International Atomic Energy Agency, Vienna).

IAEA (1985). International Atomic Energy Agency, *Assessment of Radioactive Contamination in Man 1984*, IAEA Publication No. STI/PUB/674 (International Atomic Energy Agency, Vienna).

ICRP (1960). International Commission on Radiological Protection, "Report of ICRP Committee II on Permissible Dose for Internal Radiation (1959), with Bibliography for Biological, Mathematical and Physical Data," Health Phys. **3**.

ICRP (1968). International Commission on Radiological Protection, *Report of Committee IV on Evaluation of Radiation Doses to Body Tissues from Internal Contamination Due to Occupational Exposure*, ICRP Publication 10 (Pergamon Press, New York).

ICRP (1971). International Commission on Radiological Protection, *Report of Committee IV on the Assessment of Internal Contamination Resulting from Recurrent or Prolonged Uptakes*, ICRP Publication 10A (Pergamon Press, New York, New York).

ICRP (1975). International Commission on Radiological Protection, *Report of the Task Group on Reference Man*, ICRP Publication 23 (Pergamon Press, New York, New York).

ICRP (1977). International Commission on Radiological Protection, *Recommendations of the International Commission on Radiological Protection*, ICRP Publication 26 (Pergamon Press, New York, New York).

ICRP (1979). International Commission on Radiological Protection, *Limits for Intakes of Radionuclides by Workers*, ICRP Publication 30, Part 1, Ann.

of the ICRP **2**, No. 3/4.
ICRP (1982). International Commission on Radiological Protection, *General Principles of Monitoring for Radiation Protection of Workers*, ICRP Publication 35 (Pergamon Press, New York, New York).
JACKSON, S. (1966). "Creatinine in urine as an index of urinary excretion rate," Health Phys. **12**, 843.
JACKSON, S. AND DOLPHIN, G. W. (1966). "The estimation of internal radiation dose from metabolic and urinary excretion data for a number of important radionuclides," Health Phys. **12**, 481.
JOHNSON, J.R. (1984). "Frequency of Bioassay Monitoring for Internal Contamination and Its Relationship to Sensitivity and Accuracy Requirements," Health Physics **46**, 715.
JOHNSON, J. R. (1985). "Internal dosimetry for radiation protection," page 369 in *The Dosimetry of Ionizing Radiation*, Vol. 1, Kase, K. R., Bjärngard, B. E., and Attix, F. H., Eds. (Academic Press, New York, New York).
KANAPILLY, G. M., RAABE, O. G., GOH, C. H. T. AND CHIMENTI, R. A. (1973). "Measurement of *in vitro* dissolution of aerosol particles for comparison to *in vivo* dissolution in the lower respiratory tract after inhalation," Health Phys. **24**, 497.
KANAPILLY, G. M., RAABE, O. G. AND BOYD, H. A. (1974). "A Method for determining the dissolution characteristics of accidentally released radioactive aerosols," page 1237 in *Proceedings of the Third International Congress of the Radiation Protection Association*, Washington, D.C., September 9–14, 1973, Report No. CONF-730907 (U.S. Atomic Energy Commission, Washington, D.C.).
KEANE, A. T. AND BREWSTER, D. R. (1983). "Calibration of a decay-product collection and counting apparatus for the determination of exhaled thoron," Health Phys. **45**, 801.
LEGGET, R. W. (1986). "Predicting the retention of Cs in individuals," Health Phys. **50**, 747.
LUCAS, H. F. (1957). "Improved low-level alpha scintillation counter for radon", Rev. Sci. Instrum. **28**, 690.
LUCAS, H. F., KUCHTA, D. R., SHA, J. Y., AND MARKUM, F. (1971). "Comparison of results of breath analyses at MIT and ANL," page 152 in *Radiological Physics Division Annual Report*, July, 1970–June, 1971, Report ANL-7860, Part II. (Argonne National Laboratory, Argonne, Illinois).
MASSE, F. X. AND BOLTON, M. M. (1970). "Experience with a low cost chair-type detector system for the determination of radioactive body burdens of M.I.T. radiation workers," Health Phys. **19**, 27.
MCCLELLAN, R. O., BOECKER, B.B., JONES, R. K., BARNES, J. E., CHIFFELLE, T. L., HOBBS, C. H. AND REDMAN, H. C. (1972). "Toxicity of inhaled radiostrontium in experimental animals," page 149 in *Biomedical Implications of Radiostrontium Exposure*, Goldman, M. and Bustad, L. K., Eds., Report No. CONF-710201 (U.S. Atomic Energy Commission, Washington, D.C.).
MORROW, P. E., LEACH, L. J., SMITH, F.A., GELEIN, R. M., SCOTT, J. B., BEITER, H. D. AMATO, F. J., PICANO, J. J., YUILE, C. L. AND CONSLER, T. G. (1982). *Metabolic Fate and Evaluation of Injury in Rats and Dogs*

Following Exposure to the Hydrolysis Products of Uranium Hexafluoride, Report NUREG/CR-2268 (National Technical Information Service, Springfield, Virginia).

MOSS, O. R. AND KANAPILLY, G. M. (1980). "Dissolution of inhaled aerosols." page 105 in *Generation of Aerosols and Facilities for Exposure Experiments*, K. Willeke, Ed. (Ann Arbor Science, Ann Arbor, Michigan).

NCRP (1959). National Council on Radiation Protection, *Maximum Permissible Body Burdens and Maximum Permissible Concentrations of Radionuclides in Air and in Water for Occupational Exposure*, NCRP Report No. 22 including Appendix 1 issued in 1963 (National Council on Radiation Protection and Measurements, Bethesda, Maryland).

NCRP (1971). National Council on Radiation Protection and Measurements, *Basic Radiation Protection Criteria*, NCRP Report No. 39 (National Council on Radiation Protection and Measurements, Bethesda, Maryland)

NCRP (1978). National Council on Radiation Protection and Measurements, *Operational Radiation Safety Program*, NCRP Report No. 59 (National Council on Radiation Protection and Measurements, Bethesda, Maryland).

NCRP (1980). National Council on Radiation Protection and Measurements, *Management of Persons Accidentally Contaminated with Radionuclides*, NCRP Report No. 65 (National Council on Radiation Protection and Measurements, Bethesda, Maryland).

NCRP (1985). National Council on Radiation Protection and Measurements, *General Concepts for the Dosimetry of Internally Deposited Radionuclides*, NCRP Report No. 84 (National Council on Radiation Protection and Measurements, Bethesda, Maryland).

NIOSH (1974). National Institute for Occupational Safety and Health, *Criteria for a Recommended Standard-Occupational Exposure to Benzene*, HEW Publication (NIOSH) 74-137 (U.S. Government Printing Office, Washington, D.C.).

OLSON, D. G. (1972). "In vivo determination of ^{90}Sr by analyzing bremsstrahlung from the skull," page 689 in *Health Physics Operational Monitoring*, Willis, C. A. and Handloser, J. S., Eds. (Gordon and Breach, New York, New York).

PALMER, H. E. AND RIEKSTS, G. A. (1983). "1976 Hanford americium exposure incident: *in vivo* measurements," Health Phys. **45,** 893.

PALMER, H. E. AND ROESCH, W. C. (1965). "A shadow shield whole-body counter," Health Phys. **11,** 1213.

PALMER, H. E., BRANSON, B. M., COHN, S. H., DEAN, P. N., ECKART, J. A., LLOYD, R. D. AND MILLER, C. E. (1976). "Standard field methods for determining ^{131}I *in-vivo*," Health Phys. **30,** 113.

POCHIN, E. E. (1968). "Bases for the detection and measurement of deposited radionuclides," page 127 in *Diagnosis and Treatment of Deposited Radionuclides*, Kornberg, H. A. and Norwood, W. D., Eds. (Excerpta Medica Foundation, Amsterdam).

POMROY, C. AND MEASURES, M. (1979). "External gamma counting and bioassay measurements for ^{210}Pb on retired uranium miners," Health Phys. **37,** 409.

REIZENSTEIN, P. (1973). *Clinical Whole Body Counting* (John Wright and

Sons Limited, Bristol, Great Britain).
ROBINSON, B., HEID, K. R., ALDRIDGE, T. L. AND GLENN, R. D. (1983), "1976 Hanford americium exposure incident: organ burden and radiation dose estimates," Health Phys. **45**, 911.
SAVIGNAC, N. F. AND SCHIAGER, K. J. (1974). "Uranium miner bioassay systems: lead-210 in whiskers," Health Phys. **26**, 555.
SCHIAGER, K. J. AND SAVIGNAC, N. F. (1972). *Radiation Monitoring of Uranium Miners: A Comparison of Bioassay, TLD, and the Kusuetz Determinations of Current Exposures*, U.S. AEC Report C00-1500-21 (U.S. Atomic Energy Commission, Washington, D.C.).
SCHMIER, H. AND KISTNER, G. (1972). "Proposed Selected Criteria for Monitoring the Incorporation of Radioactive Substances in Occupationally Exposed Persons," page 59 in *Assessment of Radioactive Contamination in Man*, IAEA Publication No. STI/PUB/290 (International Atomic Energy Agency, Vienna).
SCHULTE, H. F. (1975). "Plutonium: Assessment of the occupational environment," Health Phys. **29**, 613.
SNYDER, W. S., FORD, M. R., MUIR, J. R. AND WARNER, G. G. (1972). "Fluctuations of daily excretion of plutonium and their interpretation for estimation of body burden," page 711 in *Health Physics Operational Monitoring*, Willis, C. A. and Handloser, J. S., Eds. (Gordon and Breach, New York, New York).
SWINTH, K. L., PARK, J. F., VOELZ, G. L. AND EWINS, J. H. (1974). "In vivo detection of plutonium in the tracheobronchial lymph nodes with a fiberoptic coupled scintillator," page 59 in *Radiation and the Lymphatic System*, Ballou, J. E., Ed. Report No. CONF-740930 (U.S. Energy Research and Development Administration, Washington, D.C.).
THOMAS, R. G. (1968). "Transport of relatively insoluble materials from lung to lymph nodes," Health Phys. **14**, 111.
THIND, K. S. (1986). "Determination of particle size for airborne UO_2 dust at a fuel fabrication work station and its implication on the derivation and use of ICRP Publication 30 derived air concentration values," Health Phys. **51**, 97.
TOOHEY, R. E., KEANE, A. T. AND RUNDO, J. (1983). "Measurement techniques for radium and the actinides in man at the Center for Human Radiobiology," Health Phys. **44**, Supplement No. 1, 323.
USNRC (1974). U.S. Nuclear Regulatory Commission, *Application of Bioassay for Uranium*, Regulatory Guide 8.11 (U.S. Nuclear Regulatory Commission, Washington, D.C.).
USNRC (1978). U.S. Nuclear Regulatory Commission, *Bioassay at Uranium Mills* (for comment), Regulatory Guide 8.22 (U.S. Nuclear Regulatory Commission, Washington, D.C.).
USNRC (1979). U.S. Nuclear Regulatory Commission, *Applications of Bioassay for I-125 and I-131*, Regulatory Guide 8.20 (U.S. Nuclear Regulatory Commission, Washington, D.C.)
USNRC (1980). U.S. Nuclear Regulatory Commission, *Applications of Bioassay*

for Fission and Activation Products, Regulatory Guide 8.26 (U.S. Nuclear Regulatory Commission, Washington, D.C.).

USNRC (1983). U.S. Regulatory Commission, *Applications of Bioassay for Tritium*, Regulatory Guide of 713-4 (U.S. Nuclear Regulatory Commission, Washington, D.C.).

VENNART, J., MAYCOCK, G., GODFREY, B. E. AND DAVIES, B. L. (1964). "Measurements of Radium in Radium Luminizers," page 277 in *Assessment of Radioactivity in Man*, Vol. II, IAEA Publication No. STI/PUB/84 (International Atomic Energy Agency, Vienna).

WHILLANS, D. W. AND JOHNSON, J. R. (1985). "Interpretation of urinary excretion rate data in the assessment of uptakes of carbon-14," page 525 in *Assessment Radioactive Contamination in Man 1984*, IAEA Publication No. STI/PUB/674. (International Atomic Energy Agency, Vienna).

WILLIAMS, R. J. (1956). *Biochemical Individuality—The Basis for the Genetotrophic Concept* (University of Texas Press, Austin).

WRENN, M. E., COHEN, N., ROSEN, J. C., EISENBUD, M. AND BLANCHARD, R. L. (1972). "In-vivo measurements of lead-210 in man," page 145 in *Assessment of Radioactive Contamination in Man*, IAEA Publication No. STI/PUB/290 (International Atomic Energy Agency, Vienna).

The NCRP

The National Council on Radiation Protection and Measurements is a nonprofit corporation chartered by Congress in 1964 to:
1. Collect, analyze, develop, and disseminate in the public interest information and recommendations about (a) protection against radiation and (b) radiation measurements, quantities, and units, particularly those concerned with radiation protection;
2. Provide a means by which organizations concerned with the scientific and related aspects of radiation protection and of radiation quantities, units, and measurements may cooperate for effective utilization of their combined resources, and to stimulate the work of such organizations;
3. Develop basic concepts about radiation quantities, units, and measurements, about the application of these concepts, and about radiation protection;
4. Cooperate with the International Commission on Radiological Protection, the International Commission on Radiation Units and Measurements, and other national and international organizations, governmental and private, concerned with radiation quantities, units, and measurements and with radiation protection.

The Council is the successor to the unincorporated association of scientists known as the National Committee on Radiation Protection and Measurements and was formed to carry on the work begun by the Committee.

The Council is made up of the members and the participants who serve on the eighty-two scientific committees of the Council. The scientific committees, composed of experts having detailed knowledge and competence in the particular area of the committee's interest, draft proposed recommendations. These are then submitted to the full membership of the Council for careful review and approval before being published.

The following comprise the current officers and membership of the Council:

Officers

President	WARREN K. SINCLAIR
Vice President	S. JAMES ADELSTEIN
Secretary and Treasurer	W. ROGER NEY
Assistant Secretary	CARL D. HOBELMAN
Assistant Treasurer	JAMES F. BERG

Members

Seymour Abrahamson
S. James Adelstein
Roy E. Albert
Peter R. Almond
Edward L. Alpen
John A. Auxier
William J. Bair
John D. Boice, Jr.
Robert L. Brent
Antone Brooks
Thomas F. Budinger
Melvin W. Carter
George W. Casarett
Randall S. Caswell
Fred T. Cross
Gerald D. Dodd
Patricia W. Durbin
Joe A. Elder
Mortimer M. Elkind
Thomas S. Ely
Edward R. Epp
Jacob I. Fabrikant
R. J. Michael Fry
Robert A. Goepp
Robert O. Gorson
Arthur W. Guy
Eric J. Hall
Naomi H. Harley
John W. Healy
William R. Hendee
Donald G. Jacobs
A. Everette James, Jr.
Bernd Kahn
Charles E. Land

George R. Leopold
Ray D. Lloyd
Arthur C. Lucas
Charles W. Mays
Roger O. McClellan
James McLaughlin
Barbara J. McNeil
Thomas F. Meaney
Charles B. Meinhold
Mortimer L. Mendelsohn
Fred A. Mettler
William E. Mills
Dade W. Moeller
A. Alan Moghissi
Robert D. Moseley, Jr.
Wesley Nyborg
Mary Ellen O'Connor
Frank L. Parker
Andrew K. Poznanski
Norman C. Rasmussen
Wiliam C. Reinig
Chester R. Richmond
James S. Robertson
Lawrence N. Rothenberg
Leonard A. Sagan
William J. Schull
Glenn E. Sheline
Roy E. Shore
Warren K. Sinclair
Lewis V. Spencer
William L. Templeton
J. W. Thiessen
Roy C. Thompson
John E. Till
Arthur C. Upton
George L. Voelz
Edward W. Webster
George M. Wilkening
II. Rodney Withers

Honorary Members

Lauriston S. Taylor, *Honorary President*

Edgar C. Barnes
Victor P. Bond
Reynold F. Brown
Austin M. Brues
Frederick P. Cowan
James F. Crow
Merrill Eisenbud
Robley D. Evans

Richard F. Foster
Hymer L. Friedell
John H. Harley
Louis H. Hempelmann, Jr.
Paul C. Hodges
George V. LeRoy
Wilfrid B. Mann

Karl Z. Morgan
Robert J. Nelsen
Harald H. Rossi
William G. Russell
John H. Rust
Eugene L. Saenger
J. Newell Stannard
Harold O. Wyckoff

68 / THE NCRP

Currently, the following subgroups are actively engaged in formulating recommendations:

SC-1: Basic Radiation Protection Criteria
SC-3: Medical X-Ray, Electron Beam and Gamma-Ray Protection for Energies Up to 50 MeV (Equipment Performance and Use)
SC-16: X-Ray Protection in Dental Offices
SC-18: Standards and Measurements of Radioactivity for Radiological Use
SC-28: Radiation Exposure from Consumer Products
SC-40: Biological Aspects of Radiation Protection Criteria
    Task Group on Atomic Bomb Survivor Dosimetry
    Subgroup on Biological Aspects of Dosimetry of Atomic Bomb Survivors
SC-43: Natural Background Radiation
SC-44: Radiation Associated with Medical Examinations
SC-45: Radiation Received by Radiation Employees
SC-46: Operational Radiation Safety
    Task Group 2 on Uranium Mining and Milling—Radiation Safety Programs
    Task Group 3 on ALARA for Occupationally Exposed Individuals in Clinical Radiology
    Task Group 4 on Calibration of Instrumentation
    Task Group 5 on Maintaining Radiation Protection Records
    Task Group 6 on Radiation Protection for Allied Health Personnel
    Task Group 7 on Emergency Planning
    Task Group 8 on Radiation Protection Design Guidelines for Particle Accelerators
    Task Group 9 on ALARA at Nuclear Power Plants
SC-47: Instrumentation for the Determination of Dose Equivalent
SC-48: Assessment of Exposure of the Population
SC-52: Conceptual Basis of Calculations of Dose Distributions
SC-57: Internal Emitter Standards
    Task Group 2 on Respiratory Tract Model
    Task Group 5 on Gastrointestinal Tract Models
    Task Group 6 on Bone Problems
    Task Group 8 on Leukemia Risk
    Task Group 9 on Lung Cancer Risk
    Task Group 10 on Liver Cancer Risk
    Task Group 11 on Genetic Risk
    Task Group 12 on Strontium
    Task Group 13 on Neptunium
    Task Group 14 on Placental Transfer
    Task Group 15 on Uranium
SC-59: Human Radiation Exposure Experience
SC-61: Radon Measurements
SC-63: Radiation Exposure Control in a Nuclear Emergency
    Task Group on Public Knowledge About Radiation
    Task Group on Exposure Criteria for Specialized Categories of the Public
SC-64: Radionuclides in the Environment
    Task Group 5 on Public Exposure from Nuclear Power
    Task Group 6 on Screening Models
    Task Group 7 on Contaminated Soil as a Source of Radiation Exposure
    Task Group 8 on Ocean Disposal of Radioactive Waste

THE NCRP / 69

             Task Group 9 on Biological Effects on Aquatic Organisms
             Task Group 10 on Low Level Waste
             Task Group 11 on Xenon
SC-65: Quality Assurance and Accuracy in Radiation Protection Measurements
SC-66: Biological Effects and Exposure Criteria for Ultrasound
SC-67: Biological Effects of Magnetic Fields
SC-68: Microprocessors in Dosimetry
SC-69: Efficacy of Radiographic Procedures
SC-70: Quality Assurance and Measurement in Diagnostic Radiology
SC-71: Radiation Exposure and Potentially Related Injury
SC-74: Radiation Received in the Decontamination of Nuclear Facilities
SC-75: Guidance on Radiation Received in Space Activities
SC-76: Effects of Radiation on the Embryo-Fetus
SC-77: Guidance on Occupational and Public Exposure Resulting from Diagnostic Nuclear Medicine Procedures
SC-78: Practical Guidance on the Evaluation of Human Exposures to Radiofrequency Radiation
SC-79: Extremely Low-Frequency Electric and Magnetic Fields
SC-80: Radiation Biology of the Skin (Beta-Ray Dosimetry)
SC-81: Assessment of Exposure from Therapy
SC-82: Control of Indoor Radon

Study Group on Comparative Risk
   Task Group on Comparative Carcinogenicity of Pollutant Chemicals
Ad Hoc Group on Medical Evaluation of Radiation Workers
Ad Hoc Group on Video Display Terminals
Task Force on Occupational Exposure Levels

In recognition of its responsibility to facilitate and stimulate cooperation among organizations concerned with the scientific and related aspects of radiation protection and measurement, the Council has created a category of NCRP Collaborating Organizations. Organizations or groups of organizations that are national or international in scope and are concerned with scientific problems involving radiation quantities, units, measurements, and effects, or radiation protection may be admitted to collaborating status by the Council. The present Collaborating Organizations with which the NCRP maintains liaison are as follows:

American Academy of Dermatology
American Association of Physicists in Medicine
American College of Nuclear Physicians
American College of Radiology
American Dental Association
American Industrial Hygiene Association
American Institute of Ultrasound in Medicine
American Insurance Association
American Medical Association
American Nuclear Society

70 / THE NCRP

American Occupational Medical Association
American Podiatric Medical Association
American Public Health Association
American Radium Society
American Roentgen Ray Society
American Society of Radiologic Technologists
American Society for Therapeutic Radiology and Oncology
Association of University Radiologists
Atomic Industrial Forum
Bioelectromagnetics Society
College of American Pathologists
Conference on Radiation Control Program Directors
Federal Communications Commission
Federal Emergency Management Agency
Genetics Society of America
Health Physics Society
National Bureau of Standards
National Electrical Manufacturers Association
Radiation Research Society
Radiological Society of North America
Society of Nuclear Medicine
United States Air Force
United States Army
United States Department of Energy
United States Department of Housing and Urban Development
United States Department of Labor
United States Environmental Protection Agency
United States Navy
United States Nuclear Regulatory Commission
United States Public Health Service

The NCRP has found its relationships with these organizations to be extremely valuable to continued progress in its program.

Another aspect of the cooperative efforts of the NCRP relates to the special liaison relationships established with various governmental organizations that have an interest in radiation protection and measurements. This liaison relationship provides: (1) an opportunity for participating organizations to designate an individual to provide liaison between the organization and the NCRP; (2) that the individual designated will receive copies of draft NCRP reports (at the time that these are submitted to the members of the Council) with an invitation to comment, but not vote; and (3) that new NCRP efforts might be discussed with liaison individuals as appropriate, so that they might have an opportunity to make suggestions on new studies and related matters. The following organizations participate in the special liaison program:

Commission of the European Communities
Commisariat a l'Energie Atomique (France)

Defense Nuclear Agency
Federal Emergency Management Agency
Japan Radiation Council
National Bureau of Standards
National Radiological Protection Board (United Kingdom)
National Research Council (Canada)
Office of Science and Technology Policy
Office of Technology Assessment
United States Air Force
United States Army
United States Coast Guard
United States Department of Energy
United States Department of Health and Human Services
United States Department of Labor
United States Department of Transportation
United States Environmental Protection Agency
United States Navy
United States Nuclear Regulatory Commission

The NCRP values highly the participation of these organizations in the liaison program.

The Council's activities are made possible by the voluntary contribution of time and effort by its members and participants and the generous support of the following organizations:

Alfred P. Sloan Foundation
Alliance of American Insurers
American Academy of Dental Radiology
American Academy of Dermatology
American Association of Physicists in Medicine
American College of Nuclear Physicians
American College of Radiology
American College of Radiology Foundation
American Dental Association
American Hospital Radiology Administrators
American Industrial Hygiene Association
American Insurance Association
American Medical Association
American Nuclear Society
American Occupational Medical Association
American Osteopathic College of Radiology
American Podiatric Medical Association
American Public Health Association
American Radium Society
American Roentgen Ray Society
American Society of Radiologic Technologists
American Society for Therapeutic Radiology and Oncology
American Veterinary Medical Association
American Veterinary Radiology Society
Association of University Radiologists

72 / THE NCRP

Atomic Industrial Forum
Battelle Memorial Institute
Center for Devices and Radiological Health
College of American Pathologists
Commonwealth of Pennsylvania
Defense Nuclear Agency
Edison Electric Institute
Edward Mallinckrodt, Jr. Foundation
Electric Power Research Institute
Federal Emergency Management Agency
Florida Institute of Phosphate Research
Genetics Society of America
Health Physics Society
Institute of Nuclear Power Operations
James Picker Foundation
National Aeronautics and Space Administration
National Association of Photographic Manufacturers
National Bureau of Standards
National Cancer Institute
National Electrical Manufacturers Association
Radiation Research Society
Radiological Society of North America
Society of Nuclear Medicine
United States Department of Energy
United States Department of Labor
United States Environmental Protection Agency
United States Navy
United States Nuclear Regulatory Commission

To all of these organizations the Council expresses its profound appreciation for their support.

Initial funds for publication of NCRP reports were provided by a grant from the James Picker Foundation and for this the Council wishes to express its deep appreciation.

The NCRP seeks to promulgate information and recommendations based on leading scientific judgment on matters of radiation protection and measurement and to foster cooperation among organizations concerned with these matters. These efforts are intended to serve the public interest and the Council welcomes comments and suggestions on its reports or activities from those interested in its work.

NCRP Publications

NCRP publications are distributed by the NCRP Publications' office. Information on prices and how to order may be obtained by directing an inquiry to:

>NCRP Publications
>7910 Woodmont Ave., Suite 1016
>Bethesda, MD 20814

The currently available publications are listed below.

Proceedings of the Annual Meeting

No.	Title
1	*Perceptions of Risk,* Proceedings of the Fifteenth Annual Meeting, Held on March 14–15, 1979 (Including Taylor Lecture No. 3) (1980)
2	*Quantitative Risk in Standards Setting,* Proceedings of the Sixteenth Annual Meeting, Held on April 2–3, 1980 (Including Taylor Lecture No. 4) (1981)
3	*Critical Issues in Setting Radiation Dose Limits,* Proceedings of the Seventeenth Annual Meeting, Held on April 8–9, 1981 (Including Taylor Lecture No. 5) (1982)
4	*Radiation Protection and New Medical Diagnostic Procedures,* Proceedings of the Eighteenth Annual Meeting, Held on April 6–7, 1982 (Including Taylor Lecture No. 6) (1983)
5	*Environmental Radioactivity,* Proceedings of the Nineteenth Annual Meeting, Held on April 6–7, 1983 (Including Taylor Lecture No. 7) (1984)
6	*Some Issues Important in Developing Basic Radiation Protection Recommendations,* Proceedings of the Twentieth Annual Meeting, Held on April 4–5, 1984 (Including Taylor Lecture No. 8) (1985)
7	*Radioactive Waste,* Proceedings of the Twenty-First Annual Meeting, Held on April 3–4, 1985 (Including Taylor Lecture No. 9) (1986)

Symposium Proceedings

The Control of Exposure of the Public to Ionizing Radiation in the Event of Accident or Attack, Proceedings of a Symposium held April 27–29, 1981 (1982)

Lauriston S. Taylor Lectures

No.	Title and Author
1	*The Squares of the Natural Numbers in Radiation Protection* by Herbert M. Parker (1977)
2	*Why be Quantitative About Radiation Risk Estimates?* by Sir Edward Pochin (1978)
3	*Radiation Protection—Concepts and Trade Offs* by Hymer L. Friedell (1979) [Available also in *Perceptions of Risk*, see above]
4	*From "Quantity of Radiation" and "Dose" to "Exposure" and "Absorbed Dose"—An Historical Review* by Harold O. Wyckoff (1980) [Available also in *Quantitative Risks in Standards Setting*, see above]
5	*How Well Can We Assess Genetic Risk? Not Very* by James F. Crow (1981) [Available also in *Critical Issues in Setting Radiation Dose Limits*, see above]
6	*Ethics, Trade-offs and Medical Radiation* by Eugene L. Saenger (1982) [Available also in *Radiation Protection and New Medical Diagnostic Approaches*, see above]
7	*The Human Environment-Past, Present and Future* by Merril Eisenbud (1983) [Available also in *Environmental Radioactivity*, see above]
8	*Limitation and Assessment in Radiation Protection* by Harald H. Rossi (1984) [Available also in *Some Issues Important in Developing Basic Radiation Protection Recommendations*, see above]
9	*Truth (and Beauty) in Radiation Measurement* by John H. Harley (1985)

NCRP Commentaries

Commentary No.	Title
1	*Krypton-85 in the Atmosphere—With Specific Reference to the Public Health Significance of the Proposed Controled Release at Three Mile Island* (1980)

2	*Preliminary Evaluation of Criteria for the Disposal of Transuranic Contaminated Waste* (1982)
3	*Screening Techniques for Determining Compliance with Environmental Standards* (1986)

NCRP Reports

No.	Title
8	*Control and Removal of Radioactive Contamination in Laboratories* (1951)
9	*Recommendations for Waste Disposal of Phosphorus-32 and Iodine-131 for Medical Users* (1951)
16	*Radioactive Waste Disposal in the Ocean* (1954)
22	*Maximum Permissible Body Burdens and Maximum Permissible Concentrations of Radionuclides in Air and in Water for Occupational Exposure* (1959) [Includes Addendum 1 issued in August 1963]
23	*Measurement of Neutron Flux and Spectra for Physical and Biological Applications* (1960)
25	*Measurement of Absorbed Dose of Neutrons and Mixtures of Neutrons and Gamma Rays* (1961)
27	*Stopping Powers for Use with Cavity Chambers* (1961)
30	*Safe Handling of Radioactive Materials* (1964)
32	*Radiation Protection in Educational Institutions* (1966)
33	*Medical X-Ray and Gamma-Ray Protection for Energies Up to 10 MeV—Equipment Design and Use* (1968)
35	*Dental X-Ray Protection* (1970)
36	*Radiation Protection in Veterinary Medicine* (1970)
37	*Precautions in the Management of Patients Who Have Received Therapeutic Amounts of Radionuclides* (1970)
38	*Protection against Neutron Radiation* (1971)
39	*Basic Radiation Protection Criteria* (1971)
40	*Protection Against Radiation from Brachytherapy Sources* (1972)
41	*Specification of Gamma-Ray Brachytherapy Sources* (1974)
42	*Radiological Factors Affecting Decision-Making in a Nuclear Attack* (1974)
43	*Review of the Current State of Radiation Protection Philosophy* (1975)
44	*Krypton-85 in the Atmosphere—Accumulation, Biological Significance, and Control Technology* (1975)
45	*Natural Background Radiation in the United States* (1975)

46 *Alpha-Emitting Particles in Lungs* (1975)
47 *Tritium Measurement Techniques* (1976)
48 *Radiation Protection for Medical and Allied Health Personnel* (1976)
49 *Structural Shielding Design and Evaluation for Medical Use of X Rays and Gamma Rays of Energies Up to 10 MeV* (1976)
50 *Environmental Radiation Measurements* (1976)
51 *Radiation Protection Design Guidelines for 0.1–100 MeV Particle Accelerator Facilities* (1977)
52 *Cesium-137 From the Environment to Man: Metabolism and Dose* (1977)
53 *Review of NCRP Radiation Dose Limit for Embryo and Fetus in Occupationally Exposed Women* (1977)
54 *Medical Radiation Exposure of Pregnant and Potentially Pregnant Women* (1977)
55 *Protection of the Thyroid Gland in the Event of Releases of Radioiodine* (1977)
56 *Radiation Exposure From Consumer Products and Miscellaneous Sources* (1977)
57 *Instrumentation and Monitoring Methods for Radiation Protection* (1978)
58 *A Handbook of Radioactivity Measurements Procedures, 2nd ed.* (1985)
59 *Operational Radiation Safety Program* (1978)
60 *Physical, Chemical, and Biological Properties of Radiocerium Relevant to Radiation Protection Guidelines* (1978)
61 *Radiation Safety Training Criteria for Industrial Radiography* (1978)
62 *Tritium in the Environment* (1979)
63 *Tritium and Other Radionuclide Labeled Organic Compounds Incorporated in Genetic Material* (1979)
64 *Influence of Dose and Its Distribution in Time on Dose-Response Relationships for Low-LET Radiations* (1980)
65 *Management of Persons Accidentally Contaminated with Radionuclides* (1980)
66 *Mammography* (1980)
67 *Radiofrequency Electromagnetic Fields—Properties, Quantities and Units, Biophysical Interaction, and Measurements* (1981)
68 *Radiation Protection in Pediatric Radiology* (1981)
69 *Dosimetry of X-Ray and Gamma-Ray Beams for Radiation Therapy in the Energy Range 10 keV to 50 MeV* (1981)

70 *Nuclear Medicine-Factors Influencing the Choice and Use of Radionuclides in Diagnosis and Therapy* (1982)
71 *Operational Radiation Safety—Training* (1983)
72 *Radiation Protection and Measurement for Low Voltage Neutron Generators* (1983)
73 *Protection in Nuclear Medicine and Ultrasound Diagnostic Procedures in Children* (1983)
74 *Biological Effects of Ultrasound: Mechanisms and Clinical Implications* (1983)
75 *Iodine-129: Evaluation of Releases from Nuclear Power Generation* (1983)
76 *Radiological Assessment: Predicting the Transport, Bioaccumulation, and Uptake by Man of Radionuclides Released to the Environment* (1984)
77 *Exposures from the Uranium Series with Emphasis on Radon and its Daughters* (1984)
78 *Evaluation of Occupational and Environmental Exposures to Radon and Radon Daughters in the United States* (1984)
79 *Neutron Contamination from Medical Electron Accelerators* (1984)
80 *Induction of Thyroid Cancer by Ionizing Radiation* (1985)
81 *Carbon-14 in the Environment* (1985)
82 *SI Units in Radiation Protection and Measurements* (1985)
83 *The Experimental Basis for Absorbed Dose-Calculations in Medical uses of Radionuclides* (1985)
84 *General Concepts for the Dosimetry of Internally Deposited Radionuclides* (1985)
85 *Mammography—A User's Guide* (1986)
86 *Biological Effects and Exposure Criteria for Radiofrequency Electromagnetic Fields* (1986)
87 *Use of Bioassay Procedures for Assessment of Internal Radionuclide Deposition*

Binders for NCRP Reports are available. Two sizes make it possible to collect into small binders the "old series" of reports (NCRP Reports Nos. 8–30) and into large binders the more recent publications (NCRP Reports Nos. 32–88). Each binder will accommodate from five to seven reports. The binders carry the identification "NCRP Reports" and come with label holders which permit the user to attach labels showing the reports contained in each binder.

The following bound sets of NCRP Reports are also available:

Volume I. NCRP Reports Nos. 8, 9, 12, 16, 22

Volume II. NCRP Reports Nos. 23, 25, 27, 30
Volume III. NCRP Reports Nos. 32, 33, 35, 36, 37
Volume IV. NCRP Reports Nos. 38, 39, 40, 41
Volume V. NCRP Reports Nos. 42, 43, 44, 45, 46
Volume VI. NCRP Reports Nos. 47, 48, 49, 50, 51
Volume VII. NCRP Reports Nos. 52, 53, 54, 55, 56, 57
Volume VIII. NCRP Reports No. 58
Volume IX. NCRP Reports Nos. 59, 60, 61, 62, 63
Volume X. NCRP Reports Nos. 64, 65, 66, 67
Volume XI. NCRP Reports Nos. 68, 69, 70, 71, 72
Volume XII. NCRP Reports Nos. 73, 74, 75, 76
Volume XIII. NCRP Reports Nos. 77, 78, 79, 80
Volume XIV. NCRP Reports Nos. 81, 82, 83, 84, 85.

(Titles of the individual reports contained in each volume are given above).

The following NCRP Reports are now superseded and/or out of print:

No.	Title
1	*X-Ray Protection* (1931). [Superseded by NCRP Report No. 3]
2	*Radium Protection* (1934). [Superseded by NCRP Report No. 4]
3	*X-Ray Protection* (1936). [Superseded by NCRP Report No. 6]
4	*Radium Protection* (1938). [Superseded by NCRP Report No. 13]
5	*Safe Handling of Radioactive Luminous Compounds* (1941). [Out of Print]
6	*Medical X-Ray Protection Up to Two Million Volts* (1949). [Superseded by NCRP Report No. 18]
7	*Safe Handling of Radioactive Isotopes* (1949). [Superseded by NCRP Report No. 30]
10	*Radiological Monitoring Methods and Instruments* (1952). [Superseded by NCRP Report No. 57]
11	*Maximum Permissible Amounts of Radioisotopes in the Human Body and Maximum Permissible Concentrations in Air and Water* (1953). [Superseded by NCRP Report No. 22]
12	*Recommendations for the Disposal of Carbon-14 Wastes* (1953). [Superseded by NCRP Report No. 81]
13	*Protection Against Radiations from Radium, Cobalt-60 and*

NCRP PUBLICATIONS / 79

Cesium-137 (1954). [Superseded by NCRP Report No. 24]

14 *Protection Against Betatron—Synchrotron Radiations Up to 100 Million Electron Volts* (1954). [Superseded by NCRP Report No. 51]

15 *Safe Handling of Cadavers Containing Radioactive Isotopes* (1953). [Superseded by NCRP Report No. 21]

17 *Permissible Dose from External Sources of Ionizing Radiation* (1954) including *Maximum Permissible Exposure to Man, Addendum to National Bureau of Standards Handbook 59* (1958). [Superseded by NCRP Report No. 39]

18 *X-Ray Protection* (1955). [Superseded by NCRP Report No. 26]

19 *Regulation of Radiation Exposure by Legislative Means* (1955). [Out of Print]

20 *Protection Against Neutron Radiation Up to 30 Million Electron Volts* (1957). [Superseded by NCRP Report No. 38]

21 *Safe Handling of Bodies Containing Radioactive Isotopes* (1958). [Superseded by NCRP Report No. 37]

24 *Protection Against Radiations from Sealed Gamma Sources* (1960). [Superseded by NCRP Report Nos. 33, 34, and 40]

26 *Medical X-Ray Protection Up to Three Million Volts* (1961). [Superseded by NCRP Report Nos. 33, 34, 35, and 36]

28 *A Manual of Radioactivity Procedures* (1961). [Superseded by NCRP Report No. 58]

29 *Exposure to Radiation in an Emergency* (1962). [Superseded by NCRP Report No. 42]

31 *Shielding for High Energy Electron Accelerator Installations* (1964). [Superseded by NCRP Report No. 51]

34 *Medical X-Ray and Gamma-Ray Protection for Energies Up to 10 MeV—Structural Shielding Design and Evaluation* (1970). [Superseded by NCRP Report No. 49]

Other Documents

The following documents of the NCRP were published outside of the NCRP Reports series:

"Blood Counts, Statement of the National Committee on Radiation Protection," Radiology 63, 428 (1954)

"Statements on Maximum Permissible Dose from Television Receivers and Maximum Permissible Dose to the Skin of the Whole Body," Am. J. Roentgenol., Radium Ther. and Nucl. Med. 84, 152 (1960) and Radiology 75, 122 (1960)

X-Ray Protection Standards for Home Television Receivers, Interim Statement of the National Council on Radiation Protection and Measurements (National Council on Radiation Protection and Measurements, Washington, 1968)

Specification of Units of Natural Uranium and Natural Thorium (National Council on Radiation Protection and Measurements, Washington, 1973)

NCRP Statement on Dose Limit for Neutrons (National Council on Radiation Protection and Measurements, Washington, 1980)

Control of Air Emissions of Radionuclides (National Council on Radiation Protection and Measurements, Bethesda, Maryland, 1984)

Copies of the statements published in journals may be consulted in libraries. A limited number of copies of the remaining documents listed above are available for distribution by NCRP Publications.

INDEX

Action points
  Examples, 48
  Exposure control, 46–49
Air sampling, 14
Analysis
  Biopsy, 20, 21, 30, 31
  Blood, 20, 29
  Breath, 20, 29, 30
  Feces, 20, 28, 29
  Urine, 20, 25–28
  Hair, 20, 21
  Perspiration, 20, 21

Bioassay
  Action points, 46–49
  Definition, 1
  Exposure assessment, 7
  Exposure control, 3, 46–49
  Frequency, 13
  Guidance, 10
  Interpretation, 32–45, 54
  Measurements, 49
  Medical uses, 4–6
  Metabolic data, 3
  Models, 5, 15, 19–21, 32, 51, 52
  Necessity for, 7
  Non-radionuclide applications, 1
  Participation, 11
  Personnel evaluation, 2
  Perspective, 50–54
  Program, 10–12
  Strategy, 6
  Techniques, 19–30
  Uses, 1, see radionuclides
Body burden, 1, 20, 33–44

Chelation therapy, 5, 26
Chemical toxicity, 16, 17
Collection of Samples, 31
Contamination
  Airborne, 8
  Surface, 8
Corrective action, 48

Diagnostic evaluation, 49
  Errors, 50
  Uncertainties, 51–54
Dose equivalent, 4, 15, 46
  Cumulative, 33

Exposure
  Accidental, 4–6
  Assessment, 7–9
  Chronic, 13
  Control, 3
  Detection, 15–18
  Significant, 4–6
  Single, 17

Frequency of bioassay
  Chronic exposure, 13
  Respiratory equipment, 18
  Single exposure, 18

Internal dose assessment, 51–54
Interpretation, 32–45, 54
  ^{134}Cs, 33
  Checklist, 34
  Examples, 33
  General considerations, 32, 33
  Other radionuclides, 44, 45
In-vitro measurements, 25–28, 31, 51
In-vivo measurements, 20, 33–37, 51

Models, 5, 15, 19–21, 32, 51, 52

Radionuclides
  ^{241}Am, 24, 29, 49
  ^{14}C, 25, 30
  ^{134}Cs, 33–44, 52
  ^{137}Cs, 9, 20, 22, 26, 52
  ^{3}H, 9, 14, 25, 26, 30, 45
  ^{125}I, 24, 52
  ^{131}I, 9, 22, 52
  ^{210}Pb, 45
  ^{210}Po, 45
  ^{238}Pu, 17, 18, 19, 24, 29

^{239}Pu, 17, 18, 19, 24, 29, 45
^{226}Ra, 30
^{220}Rn, 30
^{222}Rn, 30
^{90}Sr, 20
^{228}Th, 30
Transuranics, 24
^{235}U, 24
^{238}U, 15–17, 24
Uranium, 15–17, 19, 20, 27, 46
^{133}Xe, 30
Radiotoxicity assessment, 7, 9

Reference man, 52

Techniques of Bioassay
  In-vitro, 25–28
  In-vivo, 20–25
  Models, 5, 15, 19–21, 51, 52

Therapeutic procedures, 33

Uranium, 15–17, 19, 20, 27, 46

Whole-body counters, 5, 22